THE ECONOMIC SINGULARITY

*Artificial Intelligence
and the promise of
Fully Automated
Luxury Capitalism*

BY CALUM CHACE

THIRD EDITION

CONTENTS

REVIEWS OF "THE ECONOMIC SINGULARITY"

"The advance of automation, described with great care and accuracy in this book, will almost certainly constitute the substrate within which all other technological developments – be they biomedical, environmental or something else entirely – will occur, and thus within which they should be discussed as regards their value to humanity. Read "The Economic Singularity" if you want to think intelligently about the future."

Aubrey de Grey, *CSO of SENS Research Foundation; former AI researcher*

"Following his insightful foray into the burgeoning AI revolution and associated existential risks, Calum focuses his attention on a nearer term challenge – the likelihood that intelligent machines will render much of humanity unemployable in the foreseeable future. He explores the arguments for and against this assertion and provides a measured response, acknowledging the risks associated with such a radical shift in our self identity but also outlining the potential significant benefits. Once again he proves a reliable guide through this complex yet fascinating topic."

Ben Medlock, *co-founder of Swiftkey, the best-selling app on Android*

"It's important that this book and others like it are written. Not because the future will necessarily happen exactly in the way described, but because it's important to be prepared if it does. If automation compels us to shift to a different economic organisation, we better start laying the foundations for the shift right now."

Dr Stuart Armstrong, *James Martin Research Fellow at the Future of Humanity Institute, Oxford University*

"Chace does a good job of answering the question whether robots will take our jobs. What worries me more though, a bit further down the road, once these robots have become massively intelligent, is whether they may take our lives. Chace covered this issue thoroughly in his previous book, "Surviving AI".

Prof. Dr. Hugo de Garis, *author of "The Artilect War", former director of the Artificial Brain Lab, Xiamen University, China*

"The jobs of the future don't exist today and the jobs of today will not exist in the future. Technological Singularity will change everything, but its first manifestation will come in the domain of economics, most likely in the shape of technological unemployment. Calum Chace's "The Economic Singularity" does a great job of introducing readers of all levels to the future we are about to face. Chace explains what might happen and what we can do to mitigate some of the negative consequences of machine takeover. The book covers unconditional basic income, virtual environments, and alternative types of economies among other things. Highly recommended."

Dr. Roman V. Yampolskiy, *Professor of Computer Engineering and Computer Science, Director of Cybersecurity lab, Author of Artificial Superintelligence: a Futuristic Approach*

"Unprecedented productivity gains and unlimited leisure—what could possibly go wrong? Everything, says Calum Chace, if we don't evolve a social system suited to the inevitable world of connected intelligent systems.

It's a failure of imagination to debate whether there will be jobs for humans in the automated world, Chace argues - we must look farther and ask how we will organize society when labor is not necessary to provide for the necessities of life. Find an answer, and life improves for all; without one, society collapses. Read this book to understand how social and technological forces will conspire to change the world—and the problems we need to solve to achieve the promise of the Economic Singularity."

Christopher Meyer, *author of "Blur", "Future*

Wealth", and "Standing on the Sun"

"It is interesting to listen to our own language. We say things such as "to earn a living", implying that you need to earn the privilege to be alive and to live a moderately enjoyable life. This may be looked upon as strangely in the future as we would now look back and say about slaves that they had to "earn their freedom".

Calum Chace hits the nail on the head in this extremely timely book. It is probably true that there will be new types of 'jobs' in whatever niches remain best explored by humans in the near future, but we should also consider an entirely different goal for the future.

Who was it in ages past who contributed those things we most remember, over time, as being of great value? It was they who contributed to the arts, the sciences and invention. But who were those people? Throughout the majority of history, these were mainly the people who either did not have to have a 'job' (because they were part of an aristocracy that had a different role to play while being supported by property and subjects), as well as the artists, artisans, philosophers or scientists who were directly supported by those patrons and therefore did not have the need to take a typical 'job'.

It is not the typical jobs that are celebrated as the best of humanity, and therefore it probably should not be our aim to find yet more categories of such jobs. Instead, wouldn't it be much better if a greater proportion of humanity could find the means to engage in preferred and culture-creating activity? With this in mind, it seems to me that it should be our aim to get rid of the need for jobs and employment just for the purpose of survival.

Our strategies for the future should be not about finding new salary jobs, but rather about removing the need for them, and about setting up a better and more advanced social structure. This is where looking at the challenges involved and the path to a successful alternative, as Chace does in chapter 5, is essential.

Where ideas such as a universal basic income (UBI) are concerned, it is useful to keep in mind that the world is not the US. Even if there is some initial antipathy in the US, because of associations between UBI and what might naively be labeled as 'socialist' thinking, the US will not wish to be left behind if other nations successfully implement the change. The time to dive deeply into the many issues raise in this book, to start a wider conversation about those issues, and to look creatively for the most well-balanced solutions and outcomes, is now."

Randal Koene , *founder of carboncopies.org*

"The Economic Singularity is fascinating. Calum Chace brilliantly explores the enormous opportunities, and risks, presented to humanity by the rapid advance of technology, and especially artificial intelligence. I couldn't put this book down."

Ben Goldsmith, *Menhaden Capital*

"In his fast-paced new book, Calum Chace explains the challenge facing humanity: to navigate through a dramatic transition which he christens the economic singularity. The culmination of an accelerating wave of automation by robots and AI, this transition threatens to do more than displace employees from the workforce. Unexpectedly, it threatens the end of capitalism itself, and potentially the fracturing of the human species.

Chace compellingly sets out a range of options, before sharing his assessment of the most credible and desirable outcomes, so that we can reach a shared "protopia" rather than a nightmarish "Brave New World" (or worse)."

David Wood, *chairman, London Futurists*

REVIEWS OF "ARTIFICIAL INTELLIGENCE AND THE TWO SINGULARITIES"

"Thoroughly researched and persuasive. Chace's principal argument seems to be correct: we need to prepare now for the economic singularity or face a serious disruption of our civilization."
Stuart Russell, *University of California, Berkeley*

"A brilliantly lucid guide to the current state of AI. This is no dystopian alarm but a clarion call to start thinking about how to cope – an entertaining and thoughtful roadmap."
Mark Mardell, *Presenter, BBC Radio Four*

"Provocative and timely: a highly readable, and

well-informed, discussion of potential economic changes that we should be thinking hard about *now.*"

Margaret Boden, *University of Sussex*

"Calum Chace, an extraordinary thinker and story-teller, presents a thorough, yet easy-to-digest account of the most pressing issues surrounding AI."

Irakli Beridze, *Centre for AI and Robotics, United Nations (UNICRI)*

"This book provides a valuable resource charting the history of AI and its current capabilities. Chace thoroughly explores the amazing potential of AI, but is wise to call for proactive planning to ensure a positive outcome."

Ray Eitel-Porter, *Accenture*

"This is the book I wish I had written. It provides a great introduction to AI, and challenges us all to engage proactively in the conversation we need to have about our future."

Shamus Rae, *founder and CEO of Engine B*

"This is a complete guide for anyone wanting to understand AI and its impact on our future. As always Calum synthesises complex technical and philosophical concepts, and makes them wonderfully accessible."

Daniel Hulme, *CEO, Satalia*

"I strongly recommend that you read "AI and the Two Singularities". The writing style is very clear and entertaining, and Calum's knowledge of the field is truly remarkable."

Simon Thorpe, *director, Toulouse Mind & Brain Institute*

ABOUT THE AUTHOR

Calum Chace is a best-selling author of fiction and non-fiction books, focusing on the subject of artificial intelligence. He is a regular speaker on artificial intelligence and related technologies, and runs a blog on the subject at www.pandoras-brain.com.

Before becoming a full-time writer, Calum had a 30-year career in journalism and business, in which he was a marketer, a strategy consultant and a CEO.

A long time ago, Calum studied philosophy at Oxford University, where he discovered that the science fiction he had been reading since boyhood is actually philosophy in fancy dress.

Also by Calum Chace:
Stories From 2045 (editor and contributor)
Artificial Intelligence and the Two Singularities
Our Jobless Future
Surviving AI
Pandora's Brain
The Internet Startup Bible (co-author)
The Internet Consumer Bible (co-author)

For Julia and Alex

THE ECONOMIC SINGULARITY, 3rd edition

A Three Cs book
ISBN 978-0-993-21164-5

First edition published in 2016 by Three Cs
Second edition published in 2018 by Three Cs
Third edition published in 2020 by Three Cs
Copyright © Calum Chace 2016, 2018, 2020

Cover and interior design © Rachel Lawston at Lawston Design, www.lawstondesign.com
Photography © iStockphoto.com and Shutterstock.com

All rights reserved
The right of Calum Chace to be regarded as the author of this work has been asserted by him in accordance with the Copyright, Design and Patents Act 1988

PREFACE TO THE THIRD EDITION

This edition was almost complete when the world was transformed by COVID-19. It is too soon to know the lasting impacts of this ghastly disease, and the extraordinary steps taken to tackle it. No doubt some of the comments in this book will be contradicted by the virus and its aftermath, and with the benefit of hindsight, some of them may look downright silly. Being overtaken by events is an inevitable risk of writing about the future, although it rarely happens as fast or as dramatically as this.

We don't yet know whether the economic recovery from the virus and the steps taken to thwart it will be V-shaped, U-shaped, or L-shaped. That will be determined by how long the world's economies have to remain locked down or hobbled, and by the wisdom and competence of its leaders.

If the recovery is V-shaped, the lasting impacts may be subtle rather than pronounced. If the recovery is U-shaped, and the economic depression is lasting, then the world will probably emerge much changed, and the suggestions and speculations in this book may need more radical revision. That will have to wait until the fourth edition.

Whatever happens, there will be scars. Many people have died already, and many more will follow. Every death is a tragedy, and leaves grieving families and friends in its wake. Companies will close, and jobs will be lost. A lot of wealth will be erased, and people who previously enjoyed comfortable incomes will find themselves in financial peril.

The short-term changes are impossible to forecast. Will people stop shaking hands when they meet? Will the populist wave recede as people demand more honest and more competent government, or will it prosper, as fear drives people further toward nationalism and other simplistic solutions? Will the techlash be intensified, or will our increased dependence on the internet shine a kinder light on the tech giants? Will the search for a broader corporate purpose than shareholder value continue, or will it come to be seen as an unaffordable indulgence by CEOs in the bygone boom days of 2019?

Longer-term, we can be reasonably confident of a few things. People around the world have been forced to not only experiment with remote working and remote meeting, but also to get good at it. Many of them have found it surprisingly convenient, not least because of all the travel time it saves. Group meetings, reunions of families and friends, conferences and other events will surely return, but more one-to-one conversations will be held by video. The automation of delivery and shop work will be accelerated, although technology remains the greatest barrier to adoption there. There will be a plethora of digital startups, incubated by people finding online solutions to problems they never knew they had. And the greater competence shown by governments in Asia than some of their counterparts in the West may well accelerate the shift in the global balance of power.

There are a couple of things that really ought to happen as a consequence of the COVID-19 outbreak. First, we must prepare better for the next one. This one should not have come as a surprise. Bill Gates warned us about it five years in advance. Hollywood made a pretty good movie about it (Contagion), as well as a whole lot of bad ones. It was not a black swan event.

Second, we should understand better the astonishing power of exponential growth. It was powerfully displayed in the rising death rates during the early phase of contagion. Exponential growth is driving other changes in our lives, and if we understand that, we will manage them far better.

INTRODUCTION

Artificial intelligence (AI) is humanity's most powerful technology. It is getting more powerful at an exponential rate.

For most of us, the most obvious manifestation of AI today is the smartphone. We take them for granted now, but many of us are glued to them: they bring all the world's knowledge to our fingertips, as well as angry birds and zombies. They are emphatically not just a luxury for people in developed countries: they provide clever payment systems, education, and market information which enable people in the emerging markets to compete and participate in the modern world.

The evolution of smartphones so far offers an intriguing analogy for the development of AI in the future. Nobody suggested thirty years ago that we would have powerful AIs in our pockets in the form of telephones, but now that it has happened it seems obvious. It is also entirely logical. We are highly social animals. Because we have language we can communicate complicated ideas, suggestions and instructions; we can work together in large teams and organise, produce economic surpluses, develop technologies. It's because of our unrivalled ability to communicate that we control the fate of this planet and every species on it. It wasn't and couldn't have been predicted in advance, but in hindsight what could be more logical than our most powerful technology, AI, becoming available to most of us in the form of a communication device?

Thirty years ago, we didn't know how the mobile phone market would develop. Today, we don't know how the digital disruption which is transforming so many industries will evolve over the next thirty years. We don't know for certain whether technological unemployment will be the result of the automation of jobs by AI, or whether automation will facilitate the creation of new jobs in the way it has since well before the start of the industrial revolution. What is the equivalent of the smartphone phenomenon for digital disruption and automation? Chances are, it will be something different from what most people expect today, but it will seem entirely natural and predictable in hindsight.

Making forecasts is risky, especially about the future, but this book will describe a series of formidable challenges which AI will present, alongside its enormous benefits. It will argue that we should monitor more closely the changes that are happening, and that we should develop a range of policy options to secure the best possible outcomes. The range of possible outcomes is wide, from the terrible to the wonderful, and they are not pre-determined. They will be selected partly by luck, partly by their own internal logic, but partly also by the policies embraced at all levels of society.

Individuals must prepare themselves to be as flexible as possible to meet the challenges of a fast-changing world. Organisations must try and anticipate the changes most relevant to them, and adapt their strategies and tactics accordingly. Governments must frame regulations which will encourage the better outcomes and fend off the worst ones. To some extent they must deploy the huge financial and human resources at their disposal too, although given the uncertainty about future developments which will prevail at all stages, they must be cautious about this.

Automation and superintelligence are the two forces which we can already see are likely to cause huge impacts. Many people remain skeptical about one or both of them, and of course no-one knows for sure how they will unfold. But we would be very unwise to ignore them.

The arrival of superintelligence, if and when it happens, would represent a technological singularity (usually just referred to as "the singularity"). "Singularity" is a term borrowed from maths and physics, and means a point where a variable becomes infinite, and as a result, the normal rules cease to apply.[12] This would be the most significant event in human history, bar none. Working out how to survive it is the most important challenge facing humanity this century. If we avoid the pitfalls, it will improve life in ways which are quite literally beyond our imagination. A superintelligence which recursively improved its own architecture and expanded its capabilities could very plausibly solve almost any human problem you can think of. Death could become optional and we could enjoy lives of constant bliss and excitement. If we get it wrong, extinction is not the worst possible outcome. I cover this issue in a companion book, "Surviving AI".

Well before then, automation could lead to what I call an economic singularity. An economic singularity might lead to an elite owning the means of production, and suppressing the rest of us in a dystopian technological authoritarian regime. Or it could lead to an economy of radical abundance, where nobody has to work for a living, and we are all free to have fun, stretch our minds, and develop our faculties to the full. I hope and believe that the latter is possible, but we may need to influence events. The economic singularity is the subject of this book: the likelihood of it arriving, and of it being beneficial.

Parts One and Two of this book (the first third) also serve as Parts One and Two of the companion book, "Surviving AI".

PART ONE:

AI TODAY

CHAPTER 1

WHAT IS ARTIFICIAL INTELLIGENCE?

Intelligence

Considering how often the term is used, artificial intelligence (AI) is very hard to define. We can say that it is intelligence demonstrated by a machine or by software. The problem arises when we try to explain what we mean by intelligence.

Like most words used to describe what the brain does, intelligence is hard to pin down, and there are many rival definitions. Most of them contain the notion of the ability to acquire information, and use it to achieve a goal. A succinct one comes from German academic Marcus Hutter, and Shane Legg, a co-founder of a company called DeepMind that we will hear about later. It states that "intelligence measures an agent's general ability to achieve goals in a wide range of environments."[3] An even more pithy definition is:

"goal-oriented learning behaviour".

As well as being hard to define, intelligence is also hard to measure. There are many types of information that an intelligent being might want to acquire, and many types of goals it might want to achieve.

American psychologist Howard Gardner has distinguished nine types of intelligence: linguistic, logic-mathematical, musical, spatial, bodily, interpersonal, intra-personal, existential and naturalistic[4]. Just listing them is sufficient for our purposes: we don't need to examine each one.

We all know that people vary in the type of intelligence they possess. Some are good at acquiring dry factual knowledge such as the birth dates of kings and queens, yet hopeless at using their knowledge to achieve some of their goals, like making new friends. Others struggle to learn things from books or lessons, but are quick to understand what other people want, and hence become very popular. When thinking about intelligence, a host of associated notions crowd in, such as reasoning, memory, understanding, learning and planning.

Intelligence is, of course, the distinguishing feature of humans: it is the characteristic which sets us apart from other animals and makes us more powerful than them. And we are much, much more powerful than them. Genetically, we are almost identical to chimpanzees, and our brains are not much heavier than theirs per kilo of body weight. But the difference in structure between our brains means that there are seven billion of us and only a few hundred thousand of them[5]. They have no say in their fate, which depends entirely on our decisions and actions: it is probably a blessing for them that they are not aware of that fact.

Our intelligence enables us to communicate, to share information and ideas, and to devise and execute plans of action. It also enables us to develop tools and technology. A single unarmed human is easy prey for a lion, but a group of humans working together, or a single human equipped with a rifle, can turn the tables very effectively.

Machine

The second half of our initial definition of artificial intelligence specified that the intelligence has to be demonstrated by a machine, or by software. The machine in question is usually a computer, although it could be any device created by humans – or indeed by aliens. Today's computers use processors made of silicon, but in future other materials like graphene may come into play.

("Computer" is an old word which pre-dates the invention of electronic machines by several centuries. Originally it meant a person who calculates, and in the early twentieth century companies employed thousands of clerks to spend long and tedious days performing tasks which they could have done in seconds if they had today's pocket calculators.)

Software is a set of instructions that tells electronic signals how to behave inside a machine. Whether intelligence resides in the machine or in the software is analogous to the question of whether it resides in the neurons in your brain or in the electrochemical signals that they transmit and receive. The answer is that both are required.

Narrow and general AI (ANI and AGI)

We need to discriminate between two very different types of artificial intelligence: artificial narrow intelligence (ANI) and artificial general intelligence (AGI[6]), which are also known respectively as weak AI and strong AI.

The easiest way to make this distinction is to say that an AGI is an AI system which can carry out any cognitive function that an adult human can. We have long had computers which are superintelligent in specific tasks: some can add up much better than any human, and others can play chess better than the best human chess grandmaster. But computers are a very long way from being able to beat humans at every intellectual endeavour. They are an interesting combination of clever and stupid: able to devise the most exquisitely ingenious chess move while the building around them is burning down.

One of the differences between our general intelligence and the machines' narrow intelligence is that we can learn useful lessons from one activity and apply them to another. Learning to play snakes and ladders makes it easier to learn how to play Ludo. A machine that learns snakes and ladders has to forget that in order to learn how to play Ludo. This is called "catastrophic forgetting", and it is something that AI researchers are trying to overcome.[7]

One of the fundamental distinctions between narrow AI and AGI involves goal-setting. Narrow AI does what we tell it to – even if the law of unintended consequences means that we sometimes wish that we had expressed our instructions differently. An AGI will have the ability to reflect on its goals and decide whether to adjust them. It will have volition.

Intelligence and consciousness

Intelligence, the ability to process information and solve problems, is very different to consciousness, the possession of subjective experience. We value intelligence highly, since it is the source of our power, but we value consciousness even more. Most humans are happy to kill and eat animals which we deem to have a lower level of consciousness than our own.

There is no reason to suppose that humans have attained anywhere near the maximum possible level of intelligence, and it seems highly likely that we will eventually create machines that are considerably more intelligent than us in all respects – assuming we don't blow ourselves up first. We don't yet know whether those machines will be conscious, let alone whether they will be more conscious than us – if that is even a meaningful question. Many people assume that an AGI will have self-awareness and be conscious, but this is a hunch rather than a proposition that can be proved yet.

Terminology

Some people dislike the term artificial intelligence. Pointing out that cars are not called artificial horses, and planes are not called artificial birds, they prefer terms like machine intelligence, or cognitive computing. I sympathise, but for the moment at least, the term artificial intelligence, or AI, is the one understood by the broadest range of people. I will use the terms machine intelligence and artificial intelligence as synonyms, and sometimes the term "machines" will encompass artificial intelligence systems.

CHAPTER 2

A SHORT HISTORY OF AI

Durable myths

Stories about artificially intelligent creatures go back at least as far as the ancient Greeks. Hephaestus (Vulcan to the Romans) was the blacksmith of Olympus: as well as creating Pandora, the first woman, he created life-like metal automatons.

More recently, science fiction got started with Mary Shelley's Frankenstein in the early nineteenth century, in which the eponymous doctor created both artificial life and artificial intelligence. A century later, in 1921, Karel Capek's play RUR (Rossum's Universal Robots) introduced the idea of an uprising in which robots eliminate their human creators.

Alan Turing

The brilliant British mathematician and code-breaker Alan Turing is often described as the father of both computer science and artificial intelligence. But his most famous achievement had nothing to do with either: it was breaking the German naval ciphers at the code-breaking centre at Bletchley Park during the Second World War. He used complicated machines known as bombes, which eliminated enormous numbers of incorrect solutions to the codes so as to arrive at the correct solution. His work is estimated to have shortened the war by two years, but incredibly, his reward was to be prosecuted for homosexuality. He was injected with synthetic oestrogen which rendered him impotent, and he died two years later. It took 57 years before a British government apologised for this barbaric behaviour.

Before the war, in 1936, Turing had already devised a theoretical device called a Turing machine. It consists of an infinitely long tape divided into squares, each bearing a single symbol. Operating according to the directions of an instruction table, a reader moves the tape back and forth, reading one square – and one symbol – at a time. Together with his PhD tutor Alonzo Church, he formulated the Church-Turing thesis, which says that a Turing machine can simulate the logic of any computer algorithm.

(The word algorithm comes from the name of a 9th-century Persian mathematician, Al-Khwarizmi.[8] It means a set of rules or instructions for a person or a computer to follow. It is different from a programme, which gives a computer precise, step-by-step instructions how to handle a very specific situation such as opening a spreadsheet, or calculating the sum of a column of figures.)

Turing is also famous for inventing a test for artificial consciousness which has become known as the Turing Test, in which a machine proves that it is conscious by rendering a panel of human judges unable to determine that it is not. Turing's original idea was to allow five minutes for the test, but later versions require a much longer examination. No-one has come up with a better test, but we cannot preclude the possibility that a non-conscious AI trained on a vast corpus of human conversations will one day be able to conduct such a sophisticated conversation with human judges over many hours that they would be unable to declare it non-conscious. That being said, can you be 100% certain that any other human is conscious?

The birth of computing

The first design for a Turing machine was made by Charles Babbage, a Victorian academic and inventor, long before Turing's birth. Babbage never finished the construction of his devices, although working machines based on his designs have now been built. His Difference Engine (designed in 1822) would carry out basic mathematical functions, and the Analytical Engine (design never completed) would carry out general purpose computation. It would accept as inputs the outputs of previous computations recorded on punch cards. Babbage's collaborator Ada Lovelace has been described as the world's first computer programmer thanks to some of the algorithms she created for the Analytical Engine.

The first electronic digital computer was the Colossus, built by code-breakers at Bletchley Park (although not by Turing). But the first general-purpose computer to be completed was ENIAC (Electronic Numerical Integrator And Computer), built at the Moore School of Electrical Engineering in Philadelphia, and unveiled in 1946. Like many technological advances, it was funded by the military, and one of its first applications was a feasibility study of the hydrogen bomb. While working on ENIAC's successor, EDVAC (Electronic Discrete Variable Automatic Computer), the brilliant mathematician and polymath John von Neumann wrote a paper describing an architecture for computers which remains the basis for the great majority of today's machines.

The Dartmouth Conference

The arrival of computers combined with a series of ideas about thinking by Turing and others led to the conjecture that "every … feature of intelligence can in principle be so precisely described that a machine can be made to simulate it." This was the claim of the organisers of a month-long conference at Dartmouth College in New Hampshire in the summer of 1956, which quickly became seen as the foundation event for the science of artificial intelligence. The organisers included John McCarthy, Marvin Minsky, Claude Shannon, and Nathaniel Rochester, all of whom went on to contribute enormously to the field.

In the years following the Dartmouth Conference, some impressive advances were made in AI, and some even more impressive claims were advanced for its potential. Machines were built that could solve high school maths problems, and a programme called Eliza became the world's first chatbot, occasionally fooling users into thinking that it was conscious.

These successes and many others were made possible in part by surprisingly free spending by military research bodies, notably the Defence Advanced Research Projects Agency (DARPA, originally named ARPA), which was established in 1958 by President Eisenhower as part of the shocked reaction in the US to the Soviet achievement of launching Sputnik, the first satellite to be placed into orbit around the Earth.

The optimism of the nascent AI research community in this period is encapsulated by some startling claims by its leading lights. Herbert Simon said in 1965 that "machines will be capable, within twenty years, of doing any work a man can do,"[9] and Marvin Minksy said two years later that "Within a generation ... the problem of creating 'artificial intelligence' will substantially be solved."[10] These were hugely over-optimistic claims which with hindsight look like hubris. But hindsight is a wonderful thing, and it is unfair to criticise harshly the pioneers of AI for under-estimating the difficulty of replicating the feats which the human brain is capable of.

Perceptrons and GOFAI

One of the algorithms which inspired the highest hopes in these early days was a simplified model of the neurons in the human brain, invented in 1957 by Frank Rosenblatt. Its instantiation in a machine called a Perceptron was the world's first artificial neural network (ANN). The optimism was quashed in 1969 by Perceptrons, a book published by Marvin Minsky and Seymour Papert, two of AI's founding figures. They seemed to provide mathematical proof that artificial neural networks had crippling limitations. Years later it was shown that these limitations could be overcome by more sophisticated neural networks, but at the time the critique was devastating.

For the next decade, therefore, research was focused on an approach called symbolic AI, in which researchers tried to reduce human thought to the manipulation of symbols, such as language and maths, which could be made comprehensible to computers. This was dubbed Good Old-Fashioned AI, or GOFAI.

AI winters and springs

It became apparent that AI was going to take much longer to achieve its goals than originally expected. There were rumblings of discontent among funding bodies from the late 1960s, and they crystallised in a report written in 1973 by mathematician James Lighthill for the British Science Research Council. A particular problem identified in the Lighthill report is the "combinatorial problem", whereby a simple task involving two or three variables becomes vast and possibly intractable when the number of variables is increased. Thus simple AI applications which looked impressive in laboratory settings became useless when applied to real-world situations.

From 1974 until around 1980 it was very hard for

AI researchers to obtain funding, and this period of relative inactivity became known as the first AI winter. This bust was followed in the 1980s by another boom, thanks to the advent of expert systems, and the Japanese fifth generation computer initiative. Expert systems attempt to solve narrowly-defined problems from single domains of expertise (for instance, litigation) using vast data banks. They avoid the messy complications of everyday life, and do not tackle the knotty problem of trying to inculcate common sense.

Japan proclaimed its fifth generation project as the successor to the first generation of computing (vacuum tubes), the second (transistors), the third (integrated circuits) and the fourth (microprocessors). It was an attempt to build a powerful presence for the country in the fast-growing computer industry, and also to abolish the perception, widespread at the time, that Japanese firms simply made cheap copies of Western-engineered products. The distinguishing feature of the fifth generation project was its adoption of massively parallel processing - the use of large numbers of processors performing co-ordinated calculations in parallel. Inevitably, Western countries responded by restoring their own funding for major AI projects. Britain launched the £350m Alvey project in 1983 and the following year DARPA set up a Strategic Computer Initiative.

Old hands began to fear that another bubble was forming, and they were proved right in the late 1980s when the funding dried up again. The reason (again) was the under-estimation of the difficulties of the tasks being addressed, but also the failure to predict that desktop computers and what we now call servers would overtake mainframes in speed and power, rendering very expensive legacy machines redundant.

The second AI winter thawed in the early 1990s, and AI research has been increasingly well funded since then. Some people are worried that the present excitement (and concern) about the progress in AI is merely the latest boom phase, characterised by hype and alarmism, and will shortly be followed by another damaging bust, in which thousands of AI researchers will find themselves out of a job, promising projects will be halted, and important knowledge and insights lost.

However there are reasons for AI researchers to be more sanguine this time round. AI has crossed a threshold and gone mainstream for the simple reason that it works. It is powering services which make a huge difference in people's lives, and which enable companies to make a lot of money: fairly small improvements in AI now make millions of dollars for the companies that introduce them. AI is here to stay because it is lucrative. This has happened partly because of Big Data.

Big Data

Big Data was the hot topic in business circles in the early 2010s. The term first appeared in October 1997, in an academic article by Michael Cox and David Ellsworth,[11] and was popularised by John Mashey at Silicon Graphics, a computer firm, to describe very large data sets which could yield surprising insights into a wide range of phenomena.[12] Businesses and other organisations (especially governments) have massively more data at their disposal than just a few years ago, and it takes considerable effort and ingenuity to figure out what to do with it – if anything.

Without really trying, we are generating and capturing much more data each year than the year before. It is said that 90% of all the data in existence was created in the last two years,[13] and that this will continue to be true for many years to come. The number of cameras, microphones, and sensors of all kinds deployed in the world is growing exponentially, and their quality is improving equally fast. We also leave digital footprints whenever we use social media, smartphones and credit cards. As well as generating more data, we are quickly expanding our capacity to store and analyse it. Turning Big Data into information and thence into understanding and insight is the job of algorithms – in other words, of AI.

Big Data was helpfully explored in a book of that name published back in 2013 by Oxford professor Viktor Mayer-Schönberger and Economist journalist Kenneth Cukier. Generally optimistic in tone, it offers a series of case studies of ways in which companies and governments are trawling through oceans of data looking for correlations which enable them to understand and influence the behaviour of their customers and citizens. Airlines can work out the best pricing policy for individual seats on each day before a flight. Hollywood studios can avoid making movies which will lose millions of dollars (but perhaps also avoid making original surprise hits).

Big Data has some interesting unexpected side-effects. It turns out that having more data beats having better data, if what you want is to be able to understand, predict and influence the behaviour of large numbers of people. Wags observe that in order to find a needle you need to start with a haystack. It turns out that if you find a reliable correlation then it often doesn't matter if there is a causal link between the two phenomena. We all know of cases where correlation has been mistaken for causation and ineffective or counter-productive policies have been imposed as a result. But if a correlation persists long enough it may provide decision-makers with at least a useful early warning signal. This has been described as the unreasonable effectiveness of data.

Big Data also has some negative aspects. Notoriously, government agencies like the NSA and Britain's GCHQ collect and store gargantuan amounts of data on us. They claim this is solely to prevent terrorist atrocities, but can they be trusted? They have been less than forthcoming about what information they are gathering, and why. What happens if the data falls into the hands of even less scrupulous organisations?

In fact it may not be the NSA and GCHQ that we have to worry about, as they struggle to offer machine learning experts the same salary or lifestyle as Google and the other tech giants. This may be less of a handicap for the security agencies of other countries, for example the Third Division of China's People's Liberation Army[14].

But concerns over privacy can themselves lead to

perverse outcomes. Cukier argues that the ebola epidemic of 2014-15 could have been halted faster - and many lives could have been saved - if phone records had been used to track and analyse the movement of people around West Africa. Fortunately ebola did not turn into the pandemic that many feared it would, and Cukier urges that we urgently review our priorities regarding privacy concerns before that does happen.[15] It doesn't look as though we managed to do that before the outbreak of COVID-19 in late 2019.

Big Data was an important phenomenon in its own right, but it achieved particular significance because it helped to make AI an overnight success, after 60 years of trying.

CHAPTER 3

THE BIG BANG: MACHINE LEARNING

In the last decade, the field of AI has undergone a quiet revolution. The dramatic increases in compute power and the amount of data available has enabled a branch of statistics called machine learning to be applied to AI. A subset of machine learning called deep learning has proved especially effective at tasks which were previously considered such hard problems that they were unlikely to be solved for many years to come.

Traditionally, at least in theory, scientists start with a hypothesis and then devise experiments to generate data to support or refute the hypothesis. Failure to adopt this approach, and simply collecting masses of data without a hypothesis to test is derided as "boiling the ocean". Machine learning, by contrast, starts with data – lots of it – and looks for patterns. Today, as mentioned, with our profusion of smartphones, sensors and tracking devices, we are generating enormous amounts of data.

The term "machine learning" is often said to have been coined by Arthur Samuel in 1959,[16] to mean the study of computer algorithms that improve automatically through experience. Machine learning algorithms use trainign data to build mathematical models in order to make predictions or take decisions without being explicitly programmed to do so.

The turning point, often called AI's Big Bang, came in 2012 when researchers in Toronto led by Geoff Hinton won an AI image recognition competition called ImageNet.[17] Hinton is a British researcher now working at Google as well as Toronto University, and the most important figure behind the rise of deep learning as the most powerful of today's AI techniques. His key colleagues at Toronto were Yann LeCun (now at Facebook), and Yoshua Bengio, a professor at the University of Montreal, and now working closely with Microsoft.

The 2012 Big Bang was the culmination of a process which began back in 1986, when Hinton realised that a process called backpropagation could train a neural network to be effective. Backpropagation retraces the steps in a neural net to identify the errors made in an initial processing of input data. As the name suggests, the error is propagated backwards down the net. It took 26 years until sufficient data and computational power was available to make backpropagation actually work.[18]

Types of machine learning

Machine learning can be supervised, unsupervised, or reinforcement learning. In supervised machine learning, the computer is given pre-labelled data, and required to work out the rules that connect them. In unsupervised learning, the machine is given no pointers, and has to identify the inputs and the outputs as well as the rules that connect them. In reinforcement learning, the system gets feedback from the environment – for instance when playing a video game, it is given the score.

Researchers are building on these basic types with variants like deep reinforcement learning,[19] and self-supervised learning. (At the risk of venturing too far into the details, supervised learning uses labels provided by the developer; unsupervised learning clusters entities, either without labels, or by inventing new labels; and self-supervised learning finds and uses labels which are already in the data.)

Machine learning researchers work with a wide range of algorithms. In his 2015 book "The Master Algorithm", Pedro Domingos categorises the researchers into five tribes: the symbolists, the evolutionaries, the Bayesians, the analogisers, and the connectionists. We have met the symbolists already: they are the purveyors of Good Old-Fashioned AI.

Evolutionists start with a group of possible solutions to a problem and produce a second "generation" by introducing small random changes and removing the least effective solutions from the population – analogous to the process of mutation and natural selection described by Darwin. Their champion from the 1960s onward was John Holland, until his death in 2015. Evolutionary computation can be very effective, but its critics argue it is circuitous and slow.

Bayesians employ a theorem developed by a Victorian British mathematician and church minister called Thomas Bayes. A Bayesian network makes hypotheses about uncertain situations and updates its degree of belief in each hypothesis according to a mathematical rule when new evidence is provided. The system generates a flow chart with arrows linking a number of boxes, each of which contains a variable or an event. It assigns probabilities to each of them happening, dependant on what happens with each of the other variables. The variables might be, for instance, missing the last train, spending a night in the open, catching pneumonia, and dying. The system tests the accuracy of the linkages and the probabilities by running large sets of data through the model, and ends up (hopefully) with a reliably predictive model.

Analogisers are the least cohesive group, according to Domingos. They rely on the observation that a new phenomenon is likely to follow the same behaviour as the previously observed phenomenon which is most like it. If you are a doctor meeting a new patient, note down her symptoms and then rifle through your files until you find the records of the patient with the most similar symptoms. The more data you have, the more accurate your analogy will be.

Deep learning

Last, and definitely not least, are the connectionists. They are trying to reverse engineer the brain by developing artificial neural networks, which have essentially been re-branded as deep learning. It has been the most successful form of machine learning so far.

Deep learning algorithms use several layers of processing, each taking data from previous layers and passing an output up to the next layer. The nature of the output may vary according to the nature of the input, which is not necessarily binary (just on or off), but can be weighted. The number of layers can vary too, with anything above ten layers seen as very deep learning.

Yann LeCun describes a typical application of deep learning as follows. "A pattern recognition system is like a black box with a camera at one end, a green light and a red light on top, and a whole bunch of knobs on the front. The learning algorithm tries to adjust the knobs so that when, say, a dog is in front of the camera, the red light turns on, and when a car is put in front of the camera, the green light turns on.

You show a dog to the machine. If the red light is bright, don't do anything. If it's dim, tweak the knobs so that the light gets brighter. If the green light turns on, tweak the knobs so that it gets dimmer. Then show a car, and tweak the knobs so that the red light gets dimmer and the green light gets brighter. If you show many examples of cars and dogs, and you keep adjusting the knobs just a little bit each time, eventually the machine will get the right answer every time.

Now, imagine a box with 500 million knobs, 1,000 light bulbs, and 10 million images to train it with. That's what a typical deep learning system is."

Deep learning researchers use different algorithms, depending on the nature of the problem they are looking to solve. Image recognition often calls for convolutional neural networks (CNNs), in which the neurons in each layer are only connected to groups of neurons in the next layer. Speech recognition more often uses recurrent networks, in which connections from each layer can loop back to an earlier layer.

The field of deep learning is evolving fast. Hinton himself has proposed replacing CNNs with capsule networks, sets of neurons which enable machines to recognise objects when they have been moved, or when the light conditions have changed.[20] There is intense debate about how much more progress can be made with deep learning, but there is no question that it has already produced astonishing advances.

Where can you see evidence of these remarkable systems? Everywhere.

CHAPTER 4

AI IS EVERYWHERE

Artificial intelligence is all around us. People in developed economies interact with AI systems many times every day without being aware of it. If it all suddenly disappeared they would notice, but its omnipresence has become unremarkable, like air.

Smartphones

The most obvious example is your smartphone. It is probably the last inanimate thing you touch before you go to sleep at night and the first thing you touch in the morning.

You may have heard the surprising fact that your smartphone has more processing power than all the computers that NASA used to send Neil Armstrong to the moon in 1969. In fact that observation is now rather old hat: apparently, a modern toaster has more processing power than the Apollo guidance computer.[21] It has been calculated that if today's iPhone could have been built back in 1957, when Frank Rosenblatt unveiled the first Perceptron, it would have cost one and a half times today's global GDP. It would have filled a 100-storey building three km long and wide, and would have used 30 times the world's current power generating capacity![22]

Your smartphone uses AI algorithms to offer predictive text and for speech recognition, and these features improve month by month. Many of the apps we download to our phones also employ AI to make themselves useful to us, and of course our phones and apps call on AI systems running elsewhere ("in the cloud") to provide us with their most powerful features. The AI in our phones becomes more powerful with each generation of phone as their processing power increases, the bandwidth of the phone networks improve, cloud storage becomes better and cheaper, and we become more relaxed (some would say negligent) about sharing enough of our personal data for the AIs to "understand" us better.

Many people in the developed economies make several internet searches a day: at the time of writing, Google carries out 82,000 searches every second[23]. Many of them are performed with the help of AI.

Logistics, recommendations...

When you visit a supermarket – or any other kind of store for that matter – the fact that the products you want are on the shelf is thanks in part to AI. The supermarkets and their suppliers are continually ingesting huge data feeds, and using algorithms to analyse them and predict what we will collectively want to buy - and when and where. The retail supply chain is enormously more efficient than it was even a decade ago thanks to these algorithms.

Other consumer-facing companies like Amazon and Netflix wear their AI on their sleeves, tempting us with products and movies based on their algorithms' analysis of what we have chosen in the past. This is the same principle as direct marketing, which has been around for decades, of course. Nowadays the data available and the tools for analysing it are much better, so people living in skyscraper apartments no longer receive junk email about lawnmowers. The marketing ecosystem is still not joined up, however. If you buy a coat online, you may be followed around the internet for a while by adverts for coats, because the ad-buying algorithm does not know you have made the purchase. Unless you are a collector of coats, these ads are a waste of the advertiser's money. That inefficiency will feel re-assuring to some, and creepy to others.

Financial services

The financial markets make extensive use of AI. High-frequency trading, where computers trade with each other at speeds no human can even follow – never mind participate in - took off in the early 21st century. The share of equity trades carried out by algorithms grew from 15% in 2000 to 80% in 2019.[24] There is still confusion about the impact of this on the financial markets. The "flash crash" of 2010, in which the Dow Jones lost almost 10% of its value in a few minutes was initially blamed on high-frequency trading, but later reports claimed that the AIs had actually mitigated the fall. The crash prompted the New York Stock Exchange to introduce "circuit breakers" which suspend trading of a stock whose price moves suspiciously quickly. The financial Armageddon which some pundits forecast has not arrived, and although there will undoubtedly be further shocks to the system, most market participants expect that new AI tools will continue to be developed for and absorbed by what has always been one of the most dynamic and aggressive sectors of the economy.

Hospitals use AI to allocate beds and other resources. Factories use robots – controlled by AI – to automate production and remove people from the most dangerous jobs. Telecoms companies, power generators and other utilities use it to manage the load on their resources.

Even though you may not see it, AI is everywhere you look. The world's biggest and most profitable companies are increasingly basing their entire existence on it.

CHAPTER 5

THE TECH GIANTS AND BE-YOND

The science of artificial intelligence is advancing rapidly, with significant steps announced almost every month. Enormous resources are being devoted to achieving these advances. Much cutting edge work in AI goes on in universities around the world, but most of it happens inside the tech giants on the US West Coast, and in China. Amazon alone spent $23bn on R&D in 2017. That's more than half of of what the UK spent on R&D spend ($45bn) – that is all the UK's companies, universities, and government bodies put together. Amazon devotes a considerable share of its R&D activity to AI – much more than the UK does.

The US tech giants are collectively known as G-MA-FIA (Google, Microsoft, Amazon, Intel, and Apple). The unfortunate acronym suits the recent backlash against technology - the "techlash". Within a bewilderingly short time, the founders and staff at the tech giants went from socially ostracised nerds to the lovable geeks of the Big Bang Theory TV show, and for a short while they were superheroes. (In case you're wondering, geeks wonder what sex in zero gravity is like; nerds wonder what sex is like.) Then it all went wrong, and now they are the tech bros; the uber-capitalist anti-heroes in a dystopian saga: too white, too male, too greedy, and too powerful. Obviously all these characterisations are caricatures, and as usual, they are all exaggerations.

At the time of writing, much of the world is in lockdown, trying to flatten the curve of the infections from COVID-19. Millions of people are experimenting with remote working and remote meeting for the first time, and communicating with their family and friends over social media and video calls instead of face-to-face in bars, living rooms, and restaurants. By the time you read this it will have become clear whether this temporary transformation in communications, often hosted or mediated by the tech giants, diluted or even reversed the techlash. Here are some brief observations on how each is dealing with AI.

The US tech giants

Google has really always been an artificial intelligence company. Back in May 2002, co-founder Larry Page said: "Google will fulfil its mission only when its search engine is AI-complete. You guys know what that means? That's artificial intelligence."25 It makes most of its huge revenues from intelligent algorithms which match adverts with readers and viewers. It is busily looking for ways to exploit its world-leading expertise in AI in new industries.

Sometimes Google enters a new industry using home-grown talent, as with its famous self-driving cars, which have now been spun off into a separate company called Waymo, and with Calico, which is looking to apply Big Data to healthcare. Other times it acquires companies with the expertise not already found inside Google, or "acqui-hires" their key talent. Its rate of acquisition reached one company a week in 2010, and by the start of 2020 it had acquired 240 of them.[26]

Most famously, it paid $500m in January 2014 for DeepMind, a two year-old company employing just 75 people which developed AIs that played video games better than people. At the time of the deal, DeepMind not only had no profits: it had no revenues. It has gone on to make astonishing contributions to the field.

In 2015, Google reorganised, establishing a holding company called Alphabet. Google, the search engine ad business, is by far the largest component by revenue, but the founders spend most of their time on the newer businesses. Most people still refer to the whole company as Google, a convention that I follow here.

Google remains at the forefront of AI research and development. We will explore DeepMind's achievements in the next chapter, and look at Waymo, and also Google's work on natural language processing in Part Two.

OpenAI is not a tech giant as such, but it is generally seen as the strongest competition to DeepMind. It was founded in 2015 by Elon Musk and Sam Altman, the founder of Y Combinator, as a non-profit with the goal of ensuring that increasingly powerful AI would remain beneficial to humanity. Musk resigned from the board in 2018 because of potential conflicts of interest with Tesla's work on AI, and the following year the company became a capped-profit company, so that it could remunerate employees better. Microsoft invested $1 billion in return for preferential access to its technologies.

Facebook lost out to Google in a competition to buy DeepMind, but in December 2013 it had hired Yann LeCun, a pioneer of deep learning. It then announced the establishment of Facebook AI Research (FAIR), with LeCun at its head. FAIR has become one of the world's most respected AI labs, responsible for major advances in image recognition systems in particular. According to Joaquin Candela, Director of Engineering for the company's Applied Machine Learning group (AML), "Facebook today *cannot exist without AI*. Every time you use Facebook or Instagram or Messenger, you may not realize it, but your experiences are being powered by AI.[27]"

For many people, the embodiment of AI today is Siri, **Apple**'s digital personal assistant that first appeared pre-loaded in the iPhone 4S in October 2011. The name is short for Sigrid, a Scandinavian name meaning both "victory" and "beauty". Apple obtained the software in April 2011 by buying Siri, the company which created it, an offshoot of a DARPA-sponsored project. Google responded a year later by launching Google Voice Search, later re-named Google Now. Microsoft has Cortana, and Amazon has Alexa, of which more below.

(In May 2016, the original creators of Siri unveiled Viv (from the Latin for "life").[28] Their company, Six Five Labs, was acquired five months later by the giant Korean company Samsung for use in its phones.)

Apple's efforts to keep up with the leaders in AI are widely thought to have been hampered by the firm's notorious culture of secrecy. ML researchers have usually spent some time in academia, and they want to publish and share their findings. Until recently, Apple forbade this, but it changed the rules in December 2016, and in July 2017 it even launched an ML blog.[29] At the time of writing, Apple's digital assistant technology was still regarded as inferior to its rivals, but it is the world's largest company by market capitalisation, and it has enormous reserves to invest in catching up.

Microsoft traditionally adopted a fast follower strategy rather than being a cutting-edge innovator. As the 21st century began, it seemed to be falling behind the newer tech giants, and was increasingly perceived as losing relevance. When Satya Nadella became CEO in February 2014 he made strenuous efforts to change this perception, and during 2016 and 2017, Microsoft executed a publicly-acknowledged pivot towards AI. Announcing a new business unit, the Microsoft AI and Research Group, comprising 5,000 staff in September 2016, Nadella said "we are infusing AI into everything we deliver".[30] A year later the company announced that the group had expanded to 8,000.[31] Microsoft is hoping that its successful Azure cloud computing service will enable it to sell its suite of AI-as-a-service tools.

In January 2013, **Amazon** bought Ivona, a Polish provider of voice recognition and text-to-speech technology. The technology was deployed in various Amazon products including Kindles, and its unsuccessful launch into mobile phones. In November 2014 the company announced Echo, a smart speaker with a digital assistant called Alexa, which helps users select music and podcasts, make to-do lists, and provides weather and other real-time information. By 2019 the company had sold 100 million Alexa-powered devices, but was still working out how to monetise them.[32]

Amazon's founder Jeff Bezos noted in a letter to shareholders that "much of what we do with machine learning happens beneath the surface. Machine learning drives our algorithms for demand forecasting, product search ranking, product and deals recommendations, merchandising placements, fraud detection, translations, and much more ... quietly but meaningfully improving core operations."

Named after the company's first CEO, **IBM's** Watson is a question answering system that ingests questions phrased in natural language, and applies knowledge representation and automated reasoning to return answers, also in natural language. It was developed in order to win the Jeopardy quiz game, which it managed in 2011 (see below), to great acclaim. Watson's architecture shows a collection of different systems and capabilities, including some employing machine learning techniques.[33]

Since then, IBM has applied the Watson brand to

a growing list of applications, notably in the medical sector. In April 2016, the company said that its cognitive computing business accounted for over a third of its $81 billion annual revenues, and was the main focus for the company's growth,[34] having invested $15 billion in the endeavour since 2010. In the decade since its Jeopardy triumph, IBM has signed partnerships with a host of public and private organisations, but making them effective has proved far harder than IBM expected, and the company has been accused of promising too much and delivering too little.[35]

No account of US artificial intelligence companies would be complete without mentioning **Intel** (Integrated Electronics), if only because its co-founder Gordon Moore is also responsible for Moore's Law, the observation and prediction that computers become twice as powerful every two years or so. As we will see below, this exponential growth is responsible for the astonishing improvement in the performance of AI. Intel was the dominant force in global chip manufacture for decades, in the ages of the mainframe and the PC. It was overtaken in chips for mobile phones and AI applications by ARM and Nvidia respectively, but it is working hard to catch up, with acquisitions like Nervana, a deep learning software provider, and Mobileye, a computer vision company with a strong presence in the self-driving car community.

China's AI giants

China has three AI giants, collectively initialised as BAT. Baidu is often described as China's Google, and

Alibaba as its Amazon. The third, Tencent, is a phone and internet giant, and is often described as China's Facebook.

Baidu is the leading search engine in China, and the fourth most visited website in the world,[36] with three-quarters of the market for internet search in China. It was founded in 2000, two years after Google, by Robin Li and Eric Wu. In May 2014 it drew attention to its ambitions for AI by hiring Andrew Ng, one of the founders of Google Brain, to head up its new AI lab in Silicon Valley, with a claimed budget of $300m over five years. Ng had been a leading figure in the development of Google Brain and then went on to help found Coursera, Stanford's online education venture. In the following two years, Baidu spent around $1.5 billion on its AI capabilities, building up a team of 1,200 researchers. Ng left Baidu again in early 2017.

Baidu has been investing heavily in self-driving vehicle technology since at least 2015. It is the leading pioneer of self-driving technology in China, with over two million km driven by the end of 2019, in 13 cities.[37]

Founded by Jack Ma in 1999, **Alibaba** overtook Walmart in 2016 to become the world's largest retailer. Its online sales exceed those of Walmart, Amazon and eBay combined. In early 2020 its market cap was $550bn, comfortably ahead of its long-time rival Tencent. Its financial services company, Ant Financial, formerly known as Alipay, is the world's largest fintech company. Because its traditional banks are relatively unsophisticated, China is widely regarded as leading the world in mobile payment systems, and Ant controls around half this market.

Alibaba's cloud computing service Aliyun (similar to Amazon Web Services) launched an AI platform called DT PAI in 2015, and in 2017 the company launched a speaker with a digital assistant called Tmall Genie, which drew comparisons with the Amazon Echo.

Tencent was founded in 1988, and its most important services are social media and games: it is the world's largest video game company. Its messaging app WeChat has a similar number of users as Facebook Messenger and WhatsApp, but a richer set of functionalities, including a very popular payments service. Half its users spend more than 90 minutes a day on the app. At the time of writing it had not succeeded in penetrating any major non-Chinese markets. Weibo is Tencent's micro-blogging site, analogous to Twitter.

Tencent established an AI lab in its home city of Shenzhen in 2016, with 50 AI researchers and 200 engineers.[38] In 2017 it announced another AI lab in Seattle, headed by senior AI researcher Yu Dong, poached from Microsoft.

Unlike in the USA, where the government has only recently begun to pay close attention to AI's potential for benefit and harm, AI is a top priority for China's government as well as its companies. AlphaGo's victories in 2016 and 2017 made a huge impact in Korea and China, where Go originated, and where it is still hugely popular. In particular, AlphaGo's defeat of Lee Sedol (see next chapter) has been described as China's "Sputnik moment", after the Soviet satellite launch in 1957 which ignited the space race, and spurred the US to launch the Apollo programme to land a man on the moon. Sputnik is also credited with inspiring DARPA's liberal expenditure on research into a range of technologies, including AI.

A White House report stated in 2016 that the US had been overtaken by China in the number of academic papers being published on AI each year,39 although papers authored in the US tend to be cited more often. In July 2017, China's State Council issued the "Next Generation Artificial Intelligence Development Plan", which called for China's AI research to be keeping pace with the most advanced labs in the world by 2020, to be making major breakthroughs by 2025, and to be the world's premier location for AI research by 2030.40

Shanghai is the focus for much of this activity, with a million square meters of the Huangpu River in its Xuhui district earmarked for AI companies. The 200-meter tall AI Tower has already been completed on the Longyao Road there, to house multinational AI companies and host exhibitions.[41]

Europe's AI giants

There are none.

Deep Mind, based in London, is probably the biggest collection of deep learning researchers in the world, and has been responsible for many impressive achievements by AI in the last few years. But it is owned by Google.

The UK and Canada rival each other for third place in the AI league table, with large numbers of talented AI researchers, many promising startups, and plenty of venture capital deployed. But they are far behind the duopoly of China and the US.

This duopoly is a problem. Europe and the rest of the world have much to contribute to our unfolding AI story. Their contribution today is certainly not zero, but it is far from what it could be. It is potentially dangerous to have two great powers directly contesting dominion of such a powerful resource. We should not be comfortable with the prospect of the Cold War between two nuclear superpowers being followed by a Code War between two AI superpowers.

Teenage sex

Google, Facebook and the other tech giants pioneered the use of machine learning, and for a while they were pretty much the only organisations with the expertise, the computing resources and the data to implement it. The joke was that machine learning was like teenage sex – everyone talks about it but no-one is actually doing it.

That is changing. In September 2015, Google announced an important change in strategy. Having built a very lucrative online advertising business based on proprietary algorithms and hardware which produced better search results than anyone else, it was open sourcing its current best AI software – a library of software for machine learning called TensorFlow.[42] The software was initially licensed for single machines only, so even very well resourced organisations weren't able to replicate the functionality that Google enjoys. In April 2016 that restriction was lifted.[43]

In October 2015, Facebook announced that it would emulate Google by open sourcing the designs for Big Sur, the server which runs the company's latest AI algorithms.[44] It later open sourced its machine learning library PyTorch. Baidu's PaddlePaddle became the first open-source deep-learning platform in China in 2016.[45]

Open sourcing confers a number of advantages for the tech giants. One is to create a level of goodwill among the AI community. Another is the fact that researchers in academia and elsewhere will learn the systems, and will be able to and inclined to work closely with Google, Facebook, and Baidu – and indeed be hired by them. Also, having more smart people working with their systems means there are more suggestions for improvement and de-bugging.

Machine learning extends beyond the tech giants

2015 was the year of the "great robot freak-out". Statements in the previous year by the "three wise men" - Stephen Hawking, Elon Musk and Bill Gates - alerted journalists to the possibility that strong AI might be developed in the foreseeable future, and be followed by superintelligence. The media responded by publishing lots of pictures of the Terminator, which was attention-grabbing, if misleading. After a pause for breath, business leaders got the idea that AI was now a technology that worked, and they began to wonder what it could do for their businesses. Indeed, they began to wonder what it would do to their industries, and whether that represented a threat.

Coca-Cola is typical of many of the world's largest companies in that around 2010 it cottoned on to the potential value of analysing the Big Data it generates about its customers, and in the last few years it has been experimenting with applications of machine learning to help with product development, and mining vast amounts of data to better understand its customers around the world.[46]

The new awareness of AI and the new availability of sophisticated machine learning tools have combined to create a thirst for information about what machine learning is and how it can be used in businesses and other organisations today. A minor industry of conferences and consultancies sprang up or morphed their offer in order to satisfy this demand. The applications of most interest to business audiences are data analytics, chatbots, and robotic process automation.

Most of the tech giants sell their products and services to consumers rather than businesses. Much of Microsoft's revenue comes from enterprises, but mostly from identical software packages. IBM is different: it generates much of its revenue from bespoke services to corporates, including consultancy – it acquired PwC's consultancy arm in 2002. As discussed above, IBM is heavily promoting AI-powered services under the Watson brand.

PwC and the other audit-based consultancies like Ernst & Young and KPMG, and other non audit-based consultancies like Accenture are also investing heavily in capabilities to help their clients deploy AI in their organisations.

The intense interest in AI has also produced a thriving ecosystem of startups. According to CB Insights, AI startups raised $17bn in 2017, $22bn in 2018, and $26bn in 2019.[47]

CHAPTER 6

HOW GOOD IS AI TODAY?

Games and quizzes: chess

The game of chess used to be thought of as one of the most challenging intellectual pursuits a person could undertake. (Because I'm rubbish at it, I still do.) It used to be thought that it would take centuries for machines to become really good at it. That was a long time ago, of course, and we are much wiser now, because as long ago as 1997, IBM's Deep Blue beat Gary Kasparov, the world's best player, in a controversial but conclusive match. Nowadays, humans have no chance against an inexpensive chess-playing programme that runs on a smartphone.

Jeopardy

IBM's next bravura AI performance came in 2011, when a system called Watson beat the best human players of the TV quiz game "Jeopardy", in which contestants are given an answer and have to deduce the question. In chapter one we noted that intelligence is not a single, unitary skill or process. Watson is an amalgam – some would say a kludge – of "more than 100 different techniques ... to analyze natural language, identify sources, find and generate hypotheses, find and score evidence, and merge and rank hypotheses." It had access to 200 million pages of information, including the full text of Wikipedia, but it was not online during the contest. The difficulty of the challenge is illustrated by the answer, "A long, tiresome speech delivered by a frothy pie topping". Contestants had to provide the question, which was "What is a meringue harangue?" Watson answered this correctly. After the game, the losing human contestant Ken Jennings famously quipped, "I for one welcome our new robot overlords."[48]

Go

In January 2016, an AI system called AlphaGo developed by Google's DeepMind beat Fan Hui, the European champion of Go, a board game. This was hailed as a major step forward: the game of chess has more possible moves (3580) than there are atoms in the visible universe, but Go has even more – 250150.[49] AlphaGo used a combination of techniques: it started by ingesting a database of 30 million moves in human-played games, and then used reinforcement learning to improve while playing many games against iterations of itself. Finally, it used a Monte Carlo search technique to select the best move in real games.

A match against perhaps the world's best-ever Go player, the Korean professional Lee Sedol, followed in March 2016. Sedol was confident, believing it would take a few more years before a computer could beat him. Most computer scientists agreed with him. He was genuinely shocked to lose the series four games to one, and observers were impressed by AlphaGo's sometimes unorthodox style of play. Famously, its move 37 in the second game looked like a terrible mistake at first, but turned out to be inspired and pivotal.

AlphaGo's achievement was another milestone in computer science, and perhaps more importantly, a breakthrough in human understanding that something important is happening, especially in the Far East, where the game of Go is far more popular than it is in the West: it was reported that 100 million people watched the games. As mentioned above, AlphaGo's achievement was a Sputnik moment for China.

AlphaGo's final contest took place a year later against the Chinese world number 1 player, Ke Jie. AlphaGo won 3-0, and afterwards, DeepMind declared it was going into retirement, with its team moving onto other projects. However, in October 2017, DeepMind unveiled AlphaGo Zero, a system with no training on previous games of Go. It was given the rules of the game and then played millions of games against itself. Within three days it was better than the version of AlphaGo which beat Lee Sedol. Three months later, DeepMind announced Alpha Zero. Within 24 hours, this system was not only better than all previous versions, but it also beat the world's best chess computer programme,

Stockfish.[50]

e-sports

In August 2017, a system developed by Elon Musk's AI startup OpenAI beat a human professional player of an online battle game called Dota 2. Although the contest was not staged in the most complex version of the game, Musk declared that the game was "vastly more complex than traditional board games like chess and Go."[51]

Not to be outdone, DeepMind announced in October 2019 that its AlphaStar system had achieved grandmaster status (better than 99.8% of all human players) at Starcraft 2, a real-time strategy game of galactic warfare. Unlike chess and Go, Starcraft players have imperfect information, and there is no definitive route to success.[52]

Self-driving vehicles

Another landmark demonstration of the power of AI began inauspiciously in 2004. DARPA offered a prize of $1 million to any group which could build a car capable of driving itself around 150-mile course in the Mojave Desert in California. The best contestant was a converted Humvee named Sandstorm which got stuck on a rock after only 7miles.[53]

This was not the first experiment with self-driving cars. From 1987 to 1995, the European Union spent $750m with Daimler Benz and others on the Prometheus project (the PROgraMme for a European Traffic of Highest Efficiency and Unprecedented Safety).[54] There were some impressive technical achievements, but ultimately the project faded. Fortunately, among other things, we have got better at devising acronyms since then.

A dozen years after the Sandstorm got stuck on its rock, Google's self-driving cars were spun out into a new company called Waymo, and by the start of 2020 they had driven well over 20 million miles without being unambiguously responsible for a single accident (up from ten million miles a year before).[55] They have been rear-ended by human drivers a few times, but this is because they obey traffic regulations scrupulously, and we humans are not used to other drivers doing that. They also drive millions of miles each day in computer simulations.

The automotive industry's initial response to the implicit challenge from Google and others was slow and piecemeal. In part this is because the car industry thinks in seven-year product cycles, while the technology industry thinks in one-year cycles at most. Most of the large car companies seemed to be convinced that self-driving technology would be introduced gradually over many years, with adaptive cruise control and assisted parking bedding in during the lifetime of one model, and assisted overtaking being introduced gradually with the next model, and so on. That was far too slow for the tech titans of Silicon Valley. Google, Tesla, Uber and others are racing towards full automation as soon as it can be safely introduced.

Today, a subsidiary of General Motors called Cruise is generally regarded as the closest follower to Waymo in the US. As of March 2020, there are at least 40 other self-driving car companies licensed to test their vehicles in the US.[56]

At present, human safety drivers are almost always present. Waymo claims that in 2018, its cars drove for 11,000 between "disengagements", when the car requires a human to take over.[57] Rivals replied that this was partly because Waymo's testing takes place in light-traffic and kindly weather conditions on the straight roads of Chandler, a suburb of Phoenix, Arizona.

Uber was one of the leading players until March 2018, when one of its cars killed a woman named Elaine Herzberg as she was crossing the road in Tempe, Arizona. Uber withdrew from testing for two years, and after an extensive investigation, Uber was found in November 2019 to be not criminally liable, but its approach to safety was criticised.[58]

Another controversial player is Tesla, co-founded and run by Elon Musk. Unlike the others, Tesla thinks that gradually increasing the amount of assistance offered to drivers is a safe path to full automation. Google was the first to abandon this approach when it found that human drivers became complacent and negligent when their involvement was only required occasionally. Tesla is also unusual in claiming that automation can be achieved without the use of a laser scanning technology called lidar (laser light + radar). On the other hand, because Tesla has so many cars on the road, it is collecting more data about cars and their environments than its rivals.

Baidu is the leading developer of self-driving cars in China, and test programmes are also under way in several European countries. In Russia, Yandex has been working on the problem since 2016.

No-one can say for certain when self-driving cars will be commonplace on our roads, but Waymo is now running some trials with no safety drivers in the vehicle. Self-driving cars will be an important and highly visible milestone in the advance of AI, and in the automation of jobs. We will cover this in more detail in chapters 10 and 18 respectively.

Search

We are strangely nostalgic about the past visions of the future, and we are often disappointed that the present is not more like the future that was foretold when we were younger. PayPal founder Peter Thiel lamented that "we were promised flying cars and instead what we got was 140 characters" (ie, Twitter, whose character limit was later raised to 280).

We haven't got flying cars (yet), but we have got something even more significant. As recently as the late 20th century, knowledge workers could spend hours each day looking for information. Today, a couple of decades after Google was incorporated in 1998, we have something close to omniscience. At the press of a button or two, you can access pretty much any knowledge that humans have ever recorded. To our great-grandparents, this would surely have been more astonishing than flying cars.

(Some people are so impressed by Google Search

that they have established a Church of Google, and offer nine proofs that Google is God, including its omnipresence, near-omniscience, potential immortality, and responses to prayer. Its slogan is "Thy domain come, Thy search be done".[59] Admittedly, at the time of writing, there are only 678 registered devotees, or "readers", at their meeting-place, a page on the internet community site Reddit.[60])

What's more, the flying cars are not so far away. Uber is working with a number of manufacturers to launch a service from city centres to airports in 2023[61], and several other well-funded companies are chasing the same goal, including Kitty Hawk, backed by Google's Larry Page and run by self-driving car pioneer Sebastian Thrun. Kitty Hawk announced a JV in February 2020 with Boeing to start a pilot service in Christchurch, New Zealand.[62]

In the early days, Google Search was achieved by indexing large amounts of the web with software agents called crawlers, or spiders. The pages were indexed by an algorithm called PageRank, which scored each web page according to how many other web pages linked to it. This algorithm, while ingenious, was not itself an example of artificial intelligence. Over time, however, Google Search has become AI-powered.

In 2013, Google announced an algorithm upgrade called Hummingbird, which used natural language processing to respond to questions such as, "what's the quickest route to Australia?" (Natural language processing is parsing sentences in ordinary language, reducing them to their basic components, processing them in order to achieve a particular result, and offering up the output in a form humans can easily understand.)

PageRank wasn't dropped, but instead became just one of the 200 or so techniques that are now deployed to provide answers. Two years later, a new technique called RankBrain was added. For searches that have not been encountered before, RankBrain converts the language into mathematical entities called vectors, which computers can analyse directly. In February 2016 Google appointed John Giannandrea, its head of AI, to run search. (Two years later he left to become a senior executive at Apple.)

Natural language processing has a great many uses. Remember spam? In the late 2000s there was talk of it crashing the internet but now you rarely see it unless you look at your junk mail box. It was tamed by machine learning and natural language processing.

Google doesn't see its main competitors in search as Microsoft's Bing, or Russia's Yandex, or even China's Baidu. 39% of purchases made online begin at Amazon, compared with 11% at Google.[63] Improving that ratio is a key aim for the search giant. The toppling of seemingly invincible industry leaders like IBM and Microsoft show how fierce and fast-moving the competition is within the technology industry. This is one of the dynamics which is pushing AI forward so fast and so unstoppably.

Image and face recognition

Deep learning has accelerated progress at tasks like

image recognition, facial recognition, natural speech recognition and machine translation faster than anyone expected. In 2012, Google announced that an assembly of 16,000 processors looking at 10 million YouTube videos had identified – without being prompted – a particular class of objects. We call them cats.[64]

Two years later, Microsoft researchers announced that their system could distinguish between the two breeds of corgi dogs.[65] (Queen Elizabeth is famously fond of corgis, so this skill would be invaluable in certain British social circles.)

We humans are very good at recognising each other's faces. Throughout history it has been vitally important to distinguish between members of your own group who will help you, and members of rival groups who may try to kill you. A Facebook AI system called DeepFace reached human-level ability to recognise human faces in March 2014, scoring 97% in a test based on a database of celebrity photos called Labelled Faces in the Wild (LFW).[66] The following year it announced the ability to recognise faces even when they are not looking towards the camera, with 83% reliability.[67]

The improvements continue: a report in 2018 found that leading systems were 20 times better than their equivalents four years previously. Face recognition technology has reached the point where Chinese consumers routinely use it to pay for goods and services. The Chinese government is extremely keen on facial recognition for surveillance, and had installed over 200 million cameras by 2019.[68]

In the West, activist citizens and governments are

more than a little skittish about the technology. San Francisco became the first city to ban its use by the police and other public agencies in May 2019, and many other jurisdictions are considering following suit. They are concerned about privacy, and also about the fact that systems are inaccurate when little training data is available for parts of the population.

Image recognition is not all about faces. The tech giants allow you to search your digital photos to find the best pictures of your family, or (of course) your cats. They are still fairly primitive, but they are reasonably reliable at identifying whether an image contains a person, a car, a ball, and so on.[69] The possible applications are endless, from saving lives by diagnosing illnesses from scans, to saving you money by identifying the brand of a shirt that you like in a video, and linking you to the retailer selling it most cheaply.

Nor is your face the only way to recognise you. Systems have been developed that can recognise you by the way you walk, dance, and type. And even at long range (200 metres and rising), by the way your heart beats.[70]

Speech recognition and translation

Speech recognition systems that exceed human performance will be available in your smartphone soon.[71] In August 2017, Microsoft announced that it had got the word-error rate of its speech transcription system down to 5.1%, the same level as humans, when dealing with quiet soundscapes, and accents that it had encountered before.[72]

Speech recognition will be highly significant when combined with translation services. International trade may boom when people around the world can understand each other almost as well as they can understand their neighbours. There will be unexpected consequences, too. Microsoft CEO Satya Nadella revealed an intriguing discovery which he called transfer learning: "If you teach it English, it learns English," he said. "Then you teach it Mandarin: it learns Mandarin, but it also becomes better at English, and quite frankly none of us know exactly why."[73]

Several companies are taking the logical step, and manufacturing earpieces which recognise the wearer's speech, translate it, and convey the result to the wearer. Douglas Adams' brilliant "Hitchhiker's Guide to the Galaxy" series first appeared on BBC radio in 1978. Very few listeners would have believed that a working version of the Babel Fish, an in-ear universal translator, and "the oddest thing in the universe"[74] would be available just 40 years later. But sure enough, in 2020, Google launches the second generation of its Pixel Buds, which promise to be exactly that.[75]

Also, machines can now beat humans at lip-reading, so the scene in the film "2001" where HAL "eavesdrops" on the astronauts while they discuss his apparent malfunction is another scenario that has stepped out of science fiction and into reality.[76]

Natural language processing (NLP)

Speech recognition is one among many components of natural language processing, whereby machines are programmed – or learn - to ingest and analyse data expressed in natural language, and generate an output, also in natural language. Early attempts to achieve this involved hand-coding sets of rules for machines to follow, but for the last few decades, statistical approaches have been preferred, and in particular, machine learning. This means that machines analyse large amounts of text, and thereby deduce the rules themselves to understand and generate language.

In 2017, researchers at Google invented the Transformer, a system which uses a function called attention to calculate the probability that a word will appear given the surrounding words. Building on this, the following year they launched BERT (Bidirectional Encoder Representations from Transformers). As the name suggests, this overcame the limitation of previous approaches, which reviewed text in one direction only.[77]

In June 2018, OpenAI published a paper on the generative pre-training (GPT) of a language model. The following February, OpenAI announced the existence of GPT-2, a greatly improved model. It did not release the full specifications for fear that bad actors might use it to generate such massive amounts of text that the internet would be flooded with fake news. The complete model was released in November.

The next generation, GPT-3, was announced in May 2020. This model has 175 billion parameters, up from 1.5 billion in its predecessor. A parameter is a calculation in a neural network that applies a weighting to data. The model's performance was impressive, but the researchers suggested that they might be reaching the limits of what could be achieved with this approach.[78]

Learning and innovating

It can no longer be said that machines do not learn, or that they cannot invent. In December 2013, DeepMind demonstrated an AI system which used unsupervised deep learning to teach itself to play old-style Atari video games like Breakout and Pong.[79] These are games which previous AI systems found hard to play because they involve hand-to-eye co-ordination.

The system was not given instructions for how to play the games well, or even told the rules and purpose of the games: it was simply rewarded when it played well and not rewarded when it played less well. As the writer Kevin Kelly noted, "they didn't teach it how to play video games, but how to learn to play the games. This is a profound difference."[80]

The system's first attempt at each game was feeble, but by playing continuously for 24 hours or so it worked out – through trial and error – the subtleties in the gameplay and scoring system, and played the games better than the best human player. What's more, when playing Breakout, the machine devised a winning strategy which its human developers had never seen before. The DeepMind system showed true general learning ability, and creativity. On seeing the demonstration, Google acquired the company for a reported $500m.

Creativity and art

In the last few years, a host of companies have been set up to create music with AI. The CEO of one of them observed in August 2017 that "a couple of years ago, AI wasn't at the stage where it could write a piece of music good enough for anyone. Now it's good enough for some use cases. It doesn't need to be better than Adele or Ed Sheeran. The aim is not 'will this get better than X?' but 'will it be useful for people?'"[81]

Machines are creating visual imagery as well as music. In June 2015, Google released pictures produced by an image recognition neural network called Deep Dream. They captured the public imagination because of their surreal, hallucinogenic properties. The network was told to look for a particular feature – for instance an eye, or a dog's head – and to modify the picture to emphasise that feature. Repeated iterations around a feedback loop created images that were sometimes beautifully haunting, and sometimes just haunting.[82]

Another creative system, called generative adversarial networks, or GANs, was developed by Google researcher Ian Goodfellow in 2014. The system contains two neural networks – one to generate a candidate image and the other to judge it, or discriminate, based on its database of prior images. Competing in this way, the system is able to produce photo-realistic images which humans cannot detect as fakes.[83]

To be clear, these systems are not conscious: they have neither experiences nor imagination. They are not emotionally affected by the images they process because they have no emotions. But creativity does not logically requires consciousness. Creativity is the use of imagination to create something original. Imagination is the faculty of having original ideas, and there seems to be no reason why that requires a conscious mind to be at work. Creativity can simply be the act of combining two existing ideas (perhaps from different domains of expertise) in a novel way.

The eminent 19th-century chemist August Kekule solved the riddle of the molecular structure of benzene while day-dreaming, gazing into a fire.[84] True, he had spent a long time before that pondering the problem, but according to his own account, his conscious mind was definitely not at work when the creative spark ignited.

Machines can be creative, but can they create art? Admittedly, this is a personal definition, and perhaps not everyone would agree, but surely art involves the application of creativity to express something of personal importance to the artist. It might be beauty, an emotion, or a profound insight into what it means to be human. (If that disqualifies a good deal of what is currently sold under the banner of art, then so be it – in fact, three cheers.)

To say something about your own experience clearly requires you to have had some experience, and that requires consciousness. Therefore, at least until a conscious artificial general intelligence (AGI) arrives, there is no art in artificial intelligence.

In April 2016, researchers from Microsoft, a Dutch university and two Dutch art galleries created an AI which analysed the way Rembrandt painted. It identified enough of his techniques and mannerisms to enable it to produce paintings in exactly his style – better than any human forger could. They had it design a new picture in Rembrandt's style, of a subject Rembrandt had never worked on, and 3D printed it to capture the old master's technique in three dimensions. Because the machine recognises patterns better than humans can, it may well teach us interesting new things about the way Rembrandt created his masterpieces. But it is not producing art.[85]

The science of what we can't yet do

A famous cartoon shows a man in a small room writing notes to stick on the wall behind him. Each note shows an intellectual task which computers are unable to carry out. On the floor there is a growing pile of discarded notes – notes which show tasks which computers now carry out better than humans. The notes on the floor include "arithmetical calculations", "play chess", "recognise faces", "compose music in the style of Bach", "play table tennis". The notes on the wall include tasks which no computer can do today, including "demonstrate common sense", but also some tasks which they now can, such as "drive cars", "translate speech in real time".

The man in the cartoon has a nervous expression: he is perturbed by the rising tide of tasks which computers can perform better than humans. Of course it may be that there will come a time the notes stop moving from the wall to the floor. Perhaps computers will never demonstrate common sense. Perhaps they will never report themselves to be conscious. Perhaps they will never decide to revise their goals. But given their startling progress to date, and the weakness of the a priori arguments that strong AI cannot be created, it seems unwise to bet too heavily on it.

It is hard to forecast what machines will and will not be able to do, and by when. It turns out to be relatively easy to programme computers to do things that we find very hard, like advanced arithmetic, but very hard to teach them how to do things that we find easy, like tying our shoelaces. This is known as Moravec's paradox, after AI pioneer Hans Moravec.[86]

We tend to forget how much progress AI has made. Although iPhones and Android phones are called "smartphones", we don't tend to think of them as instantiations of artificial intelligence. We don't tend to think of the logistical systems of the big supermarkets as examples of AI. In effect, artificial intelligence is re-defined every time a breakthrough is achieved. Computer scientist Larry Tesler pointed out that this means AI is being defined as "whatever hasn't been done yet", an observation which has become known as Tesler's Theorem, or the AI effect.

For many years, people believed that computers would never beat humans at chess. When it finally happened, it was dismissed as mere computation – mere brute force, and not proper thinking at all. The rebarbative American linguistics professor Noam Chomsky declared that a computer programme beating a human at chess was no more interesting than a bulldozer winning an Olympic gold at weight-lifting. (Maybe he was saying this before Kasparov was beaten, but if so he was very unusual.)

Computers are not (as far as we know) self-conscious. They cannot reflect rationally on their goals and adjust them. They do not (as far as we know) get excited about the prospect of achieving their goals. They do not (we believe) actually understand what they are doing when they play a game of chess, for instance.

In this sense it is fair to say that what AI systems do is "mere computation". Then again, a lot of what the human brain does is "mere computation", and it has enabled humans to achieve some wondrous things. It is not unreasonable that humans want to preserve some space for themselves at the top of the intellectual tree. But to dismiss everything that AI has achieved as not intelligent, and to conclude – as some people do – that AI research has made no progress since its early days in the 1950s and 1960s is frankly ridiculous.

Machine learning works, and in a few short years it has achieves remarkable things. But in truth, it has hardly got started. In Parts Two and Three of this book we will look at where it may take us in the coming years and decades. But first we will explore some of the problems and challenges presented by these marvels.

CHAPTER 7

AI ETHICS

Powerful new technologies offer great benefits, but they can also cause great harm. Artificial intelligence raises a host of concerns, including privacy, transparency, bias, fake news, killer robots, and algocracy. In no particular order, let's look at each of them.

Privacy

AI systems have huge appetites for data. We are generating lakes of data today, and the Internet of Things will generate oceans of it. In an intelligent environment, the past and present location of every citizen is easy to establish, along with who they have met and very possibly what they discussed. A famous cartoon in 1993 showed a dog typing away at a computer terminal with the caption, "on the internet, nobody knows you're a dog". Today, the social media platforms you use know the breed, age, and gender of your dog, and even what dog food it prefers.

Many people are understandably concerned about this information being used and mis-used by all sorts of organisations, including (depending on their political persuasion) governments, corporations, pressure groups and resourceful individuals – such as jealous spouses. As one group of activists puts it, we are increasingly transparent to organisations which are increasingly opaque to us.[87]

Perhaps we can mitigate this kind of "surveillance on steroids" with "sousveillance" - surveillance from below of those with authority over us. With cameras ubiquitous – including on drones - the actions of those in authority are constrained because they know that their actions are observed and recorded by members of the public. This is already happening with law enforcement, with police officers in the US being prosecuted for harassment in situations where they would previously have been immune from oversight. Some forces are actively embracing this development, with their officers being required to wear cameras at all times in order to pre-empt false allegations. With cameras on drones, the reach of civilian oversight can be extended so far that some are calling it "Little Brother".[88] With the watchers being watched, we may arrive at a balance called "co-veillance".[89]

The arms race over data will continue between governments, large organisations, and the rest of us. Hollywood loves the trope of the socially dysfunctional hacker who is smarter, more up-to-date and more motivated than her opposite numbers in the civil service, but perhaps we should not be comforted by that idea.

The hacker is by no means always on our side.

When forced to choose between privacy and the opportunity to share, we generally choose to share. We leave a trail of digital breadcrumbs wherever we go, both in the real world and online, and most of us are careless about it.

In part this is because many of us feel that we have nothing to worry about because we have nothing to hide. But there is a chilling effect on free speech if we start to censor ourselves because we want to stay that way. We self-censor when we are aware of the possibility that we are being surveilled, even when we know we are saying and writing nothing illegal.[90] We think twice before entering a certain term into a search engine, and we might hesitate before making friends with someone who is overtly counter-cultural.

The Chinese government is developing an unnerving demonstration of where this could lead. It is building a "social credit" database of all citizens which ranks them according to their trustworthiness. The database will incorporate all the financial and behavioural information the government can accumulate, and distil it into a single number, ranging from 350 to 950. A score above 600 qualifies you for an instant loan worth $800. At 650 you can rent a car without leaving a deposit. At 700 you are fast-tracked for a Singapore travel permit. Important jobs will require high scores.

Scores will be lowered for reprehensible shopping habits (too many video games? too much wine?) and raised for socially responsible actions, like reporting bad behaviour by others.[91] A particularly scary aspect of the system is that people receive demerits if their friends on social media are marked down.

Big Data and AI could enable governments to build apparatus of control which would make Big Brother in George Orwell's "1984" look amateurish. You're not necessarily safe from this prospect just because you live in a multi-party democracy. In April 2014, Nicole McCullough and Julia Cordray founded Peeple, an app to enable people to rate each other according to their courtesy and helpfulness. Originally conceived as a way to improve behaviours, it was widely criticised as likely to become a medium for personal attack and bullying.[92] Clearly we still have much to learn about how to conduct ourselves individually and collectively in the new world of data tsunamis and massive analytic horsepower.

Beyond the fact that we have fallen into careless habits with regard to our personal data, and we keep voting with our keystrokes to surrender ever more of it, there may be very good societal reasons to think that privacy cannot be retained, or rather restored. Technology is usually a double-edged sword, and the ability to manufacture deadly pathogens, deploy fearsome miniaturised killer drones, and unleash devastating digital viruses is becoming ever cheaper. What price privacy when the cost of a megadeath falls within the means of a gruntled teenager? Perhaps we will have no sensible option but to surrender our privacy, and rather than keeping secrets, our right would be to know what our secrets are being used for.[93]

There is also the consideration that unless we do

share our data freely, we will not reap the benefits that AI can provide. If AI systems have access to the whole population's health data, they can make the best diagnoses and prescribe the best remedies. If they know the movements of everybody in a disease-affected area, they can determine the best measures to control it. If we hold back the data, these benefits will not accrue.

These are complex issues, and will be debated over and over as new possibilities and threats open up.

Researchers are experimenting with promising approaches to squaring the circle of protecting privacy while sharing data. Training deep learning algorithms on separate data files and then sharing the outputs of the trained data can work almost as well as combining all the data into one file and using that to train the algorithm. Another technique, called homomorphic encryption, analyses encrypted data. The results can then be decrypted without the sensitive data ever having been available to the analysts in unencrypted form.[94]

Another approach to retaining or restoring privacy is to reform the internet itself. The inventor of the world wide web, Sir Tim Berners-Lee, is working on Solid, a project to decentralise the web, and give users control of the data about them, including decisions about where it is stored and who has access to it.[95] This is sometimes called web 3.0.

My data?

Here is a controversial suggestion: we may be thinking about privacy the wrong way. Activists talk about the tech giants using – or even stealing - "our data", but why do we think the data they use belongs to us? Does the act of walking into a shop (or shopping online) and looking at a product create some property – a piece of data - which you own? Or it it simply a fact about the world? What is the nature of this property, and how was it created and attributed? If every act we undertake creates a piece of property, then the world is rapidly filling up with a horribly tangled mess of property rights and the lawyers are going to get very rich.

Maybe what we should be thinking about – and indeed worrying about - is the behaviour of other people in relation to us, rather than the data that is created by our actions per se.

Imagine you walk out of a shop, and someone called Harry takes a photo of you. If there is a problem with this, surely it is more about who Harry is, and what he intends to do with the photo, rather than a fundamental, immutable issue about someone acquiring or using "your data"? Harry could be:

your friend, whom you asked to take the picture

a friend who is secretly photographing you for an element of a surprise birthday party

operating a Google Street View car's camera

carrying out market research for the retailer

a paparazzo

a private detective working for your husband's divorce lawyer

a stalker

a member of your government's intelligence services, compiling clandestine evidence for an illegal report about your entirely legitimate political activities.

This spectrum ranges from benign to downright terrifying, and there are laws covering several of the possibilities. (The laws are not always adequate, but that is another matter.) The issue is behaviour, not data.

Transparency

AI systems, and especially those using deep learning, are often described as "black boxes". They generate effective and helpful results, but we cannot see how they arrived at their answers. This is a problem when important decisions are being made about people's lives. "The computer says No" is an unacceptable response when you ask why your loan application was refused, or why your application for early release from jail was denied.

The financial services industry and the military are two sectors where this problem is particularly acute, for obvious reasons, and work is under way there as well as elsewhere to try to make deep learning systems more transparent.[96] If it succeeds, a world where many decisions are taken by AIs will be more transparent than the one we live in today, where bureaucracies are often unwilling or unable to tell us why particular decisions have been made. Humans often make decisions without being able or willing to explain their real motivations in full.

Bias

We tend to think of machines as cold, calculating,

and unemotional, which of course they are. But they are also prejudiced. Not because of any implicit beliefs of their own – they have none - but because they pick up prejudices from us. Machines learn about the meanings and associations of words from the sentences that we give them, and bias is ingrained in human thought and language. Machines build up mathematical representations of language, in which meaning is captured by numbers, or vectors, based on the other words most commonly associated with them. Sometimes this is innocuous, as when we associate flowers with positive feelings, and spiders with negative ones. But it can have unwanted implications if, for example, we read the phrase "the professor and the assistant professor", and we associate "professor" with "male", and "assistant professor" with "female".[97]

An extreme example of a biased machine was provided by Microsoft in March 2016 when it launched a chatbot called Tay on Twitter. Within 24 hours, mischievous (and perhaps in some cases vicious) humans had converted Tay into an aggressive racist. Horrified, Microsoft shut it down.

Inadvertently infecting our machines with existing bias is not necessarily making our world worse: in many cases it is simply perpetuating the problem that already exists. And at least machines will not attempt to cover up or post-rationalise their bias. If we can observe it, we can tackle it, and by tackling the bias in machines, perhaps we can surface and tackle the bias in ourselves.

Fake news and filter bubbles

A popular dystopian view of the major social media platforms is that they deliberately drive us to adopt increasingly extreme and polarised opinions, and to despise those who have the temerity to disagree with us. More and more people, it is claimed, inhabit ideologically isolated mental bubbles, shunning information sources that tolerate diverse ideas. Social media companies make money when we click links, and anger is the most effective promoter of clicks, so Twitter, Facebook and YouTube feed us a diet of increasingly splenetic invective ("clickbait") to keep us on their sites as long as possible.

Facebook and Google are twice damned in this narrative. Not only are they poisoning the well of public discourse, but they are also stealing our data for selfish commercial ends. They have invented a new form of capitalism called surveillance capitalism, which an author of a book by that name claims is no less damaging that the Spanish conquistadors.

There are elements of truth in this, but there is a lot wrong with it, too. US politics became polarised with the birth of the tea party, and that pre-dated the rise of social media by several years. The tea party was the product, not of digital media, but of the decades-long triumphal march of liberal social policies – policies which conferred tremendous benefits, but which also left large numbers of people (rightly or wrongly) feeling alienated.[98] The proximate cause was the election of Barack Obama, who took office in January 2009.

The era before social media was no paradise in which all citizens consumed a range of information sources to protect themselves from bias: most people read only one newspaper, and press barons have long been willing to lie and distort stories to push their agendas. If you're looking for the people who have dragged our politics into the gutter you should probably be looking at media owners like Rupert Murdoch (The Sun newspaper in the UK, and Fox News in the US). Fake news is much more about Fox than Facebook.

We should also remember that these things go in cycles. It is hard to deny that some of today's politicians have less integrity and less regard for truth than we have seen for a long time. But this is unlikely to be a one-way road that leads to worse and worse outcomes forever. Voters will tire of the lies, and honesty will come back into fashion at some point.

Nevertheless, the social media platforms are enormously powerful, and we are still working out out how to regulate them, both in our own personal lives, and at a societal level. In October 2019, Wikipedia founder Jimmy Wales launched a social media platform called WT Social, which is not-for-profit, in the hope that it will not succumb to the malign intentions people see in Facebook et al.[99]

Like any powerful new technology which impacts the media, AI opens up new possibilities, some of which are malign. Perhaps the most intriguing and concerning so far is deepfakes (from "deep learning" plus "fake") which replace one person's face with another (widely used in pornography, so we are told), and allow creators to place words in politicians' mouths. Some people speculate that tackling deepfakes could be the first significant use case for the blockchain, but going into detail on that would be a digression too far.

Perhaps in the future, we will have digital assistants which will tell us what impact a particular news story or tweet, etc, is having on our minds, and also how likely it is to be true. This could help combat fake news, although some would find it unacceptably algocratic (see below).

Regulators to the rescue?

We noted the backlash ("techlash") against the tech giants in chapter five. Many people are worried that they are too powerful, and argue that they need to be weakened, restrained, or controlled.

European Commission has been pondering this issue for years. It accuses the tech giants of dodging taxes by booking revenues which are transacted in cyber-space to low-tax domiciles.

The tech giants employ clever lawyers, and take full advantage of the new circumstances of online business to minimise their tax bills. Multinational companies have always presented challenges to revenue officials: in vertically integrated industries like oil and pharmaceuticals, transfer pricing arrangements between subsidiaries have long had major impacts on the income streams of governments. But it is governments which set the rules under which taxes are paid, and it is not necessarily reasonable for people or companies to pay for the mistakes which governments have made, or for their failure to keep up with new realities.

Competition legislation is necessary, and monopolies should be broken up where they form and are shown to be operating against the public interest. But companies should not be penalised just because they are big. Harm must be demonstrated, and it is hard to demonstrate the misuse of monopolistic market power to exploit consumers when products and services are provided free. There are other kinds of harm, though: the tech giants are accused of wholesale infringements of privacy, of excluding smaller competitors from markets to the detriment of consumers, and of allowing harmful content to remain on social media when it should have been removed, or should not have been uploaded in the first place.

Even here, markets often achieve what regulators cannot and should not try to. It is true that there are network effects in information-based industries which can favour the emergence of monopolies. But there is also fierce competition in the technology industry, and business models change quickly, creating losers out of winners and vice versa. The histories of IBM, Microsoft, and Apple illustrate this clearly, and Google, Amazon, Facebook are not immune.

Google makes most of its money from selling ads alongside search results, and that is vulnerable if consumers use different search engines, and if the nature of search changes. As we noted before, Google does not see the primary competition to its search engine as Bing, Microsoft's search engine, or DuckDuckGo, a search engine that promises not to retain your personal information. Its real competitor in search is Amazon.

Perhaps even more threatening to Google is the way that search will change over the coming years as we control our computers more and more by voice rather than with keyboards. When you are looking at a screen it is easy to serve you ads along with the specific information you requested. It is hard to see how this can be replicated when the machine gives you the requested information verbally.

Regulation may be required at times, but because it tends to tackle issues which have already faded, it can have modest or even negative impacts, and should therefore be embarked upon with great caution.

Killer robots

Human Rights Watch and other organisations are concerned that in the coming years, fully autonomous weapons will be available to military forces with deep pockets. They argue that lethal force should never be delegated to machines because they can never be morally responsible. Their stance has garnered a great deal of support.

In his book "Homo Deus", Yuval Hahari sets out the counter-argument with his customary genteel brutality: "Suppose two drones fight each other in the air. One drone cannot fire a shot without first receiving the go-ahead from a human operator in some bunker. The other drone is fully autonomous. Which do you think will prevail? … Even if you care more about justice than victory, you should probably opt to replace your soldiers and pilots with autonomous robots and drones. Human soldiers murder, rape and pillage, and even when they try to behave themselves, they all too often kill civilians by mistake."

Ultimately the decision whether to develop these weapons will be taken by governments and their military personnel. It is logically possible that all governments and all military commanders everywhere might refrain in perpetuity, but it is hard to believe that will happen in practice.

It is one thing for national armies to develop and stockpile these weapons, and quite another for them to become available to criminal gangs and terrorists. Between nation states, deterrence can have some meaning. For 70 years, we have had enough nuclear missiles on the planet to destroy it several times over. Not only has that never happened, but fewer people were killed in wars during those 70 years than at any time since the global population was far smaller.

The logic of deterrence is repulsive, and the world would be a safer place if it contained many fewer nuclear warheads. But the gravest danger of killer robots may come from non-state actors, as illustrated graphically in "Slaughterbots" a short film made by the Future of Life Institute (FLI), an organisation which explores and campaigns about existential risks to humanity.[100]

Algocracy

The final concern we will review here receives far less attention than the others at the moment, but could turn out to be the most significant and the most lasting of them.

In his 2006 book "Virtual Migration", Indian-American academic A. Aneesh coined the term "algocracy" for the fact that algorithms increasingly take many decisions which were formerly the responsibility of humans. They initiate and execute many of the trades on stock and commodity exchanges. They manage resources within organisations providing utilities like electricity, gas and water. They govern important parts of the supply chains which put food on supermarket shelves. This phenomenon will only increase.

As our machines get smarter, we will naturally delegate decisions to them which would seem surprising today. Imagine you walk into a bar and see two attractive people at the counter. Your eye is drawn to the blond, but your digital assistant (located now in your glasses rather than your phone) notices that and whispers to you, "hang on a minute: I've profiled them both, and the red-head is a much better match for you. You share a lot of interests. Anyway, the blond is married."

Of course many of the decisions being delegated to algorithms are ones we would not want returned to human hands – partly because the machines make the decisions so much better, and partly because the intellectual activity involved is deathly boring. It is not particularly ennobling to be responsible for the decision whether to switch a city's street lights on at 6.20 or 6.30 pm, but the decision could have a significant impact. The additional energy cost may or may not be offset by the improvement in road safety, and determining that equation could involve collating and analysing millions of data points. Much better work for a machine than a human, surely.

Other applications are much less straightforward. Take law enforcement: a company called Intrado provided an AI scoring system to the police in Fresno, California. When an emergency call names a suspect, or a house, the police can "score" the danger level of the person or the location and tailor their response accordingly.[101] Other forces use a "predictive policing" system called PredPol which forecasts the locations within a city where crime is most likely to be carried out in the coming few hours.[102] Optimists would say this is an excellent way to deploy scarce resources. Pessimists would reply that Big Brother has arrived.

The allocation of new housing stock, the best date for an important election, the cost ceiling for a powerful new drug. Arguments about which decisions should be made by machines, and which should be reserved for humans are going to become increasingly common, and increasingly vehement. Regardless whether they make better decisions than we do, not everyone is going to be content (to paraphrase Grace Jones) to be a slave to the algorithm.

Information is power. Machines may intrude on our freedom without actually making decisions. In September 2017 a research team from Stanford University was reported to have developed an AI system which could do considerably more than just recognise faces. It could tell whether their owners were straight or gay. The idea of a machine with "gaydar" is startling. It becomes shocking when you consider the uses it might be put to – in countries where homosexuals are persecuted and even prosecuted, for instance.[103] The Stanford professor who led the research later said that the technology would probably soon be able to predict with reasonable accuracy a person's IQ, their political inclination, or their predisposition towards criminality.

Has AI ethics got a bad name?

Artificial intelligence is a technology, and a very powerful one, like fire. It will become increasingly pervasive, like electricity. Like fire and electricity, AI can have positive impacts and negative impacts, and given how powerful it is and it will become, it is vital that we figure out how to promote the positive outcomes and avoid the negative outcomes.

But if AI is a technology like fire or electricity, why is the field called "AI ethics"? We don't have "fire ethics" or "electricity ethics", so why should we have AI ethics? There may be a terminological confusion here, and it could have negative consequences.

One possible downside is that people outside the field may get the impression that some sort of moral agency is being attributed to the AI, rather than to the humans who develop AI systems. The AI we have today is narrow AI: superhuman in certain narrow domains, like playing chess and Go, but useless at anything else. It makes no more sense to attribute moral agency to these systems than it does to a car, or a rock. It will probably be many years before we create an AI which can reasonably be described as a moral agent.

There is a more serious potential downside to the nomenclature. People are going to disagree about the best way to obtain the benefits of AI and minimise or eliminate its harms. That is the way it should be: science, and indeed most types of human endeavour, advance by the robust exchange of views. People and groups will have different ideas about what promotes benefit and minimises harm. These ideas should be challenged and tested against each other. But if you think your field is about ethics rather than about what is most effective, there is a danger that you start to see anyone who disagrees with you as not just mistaken, but actually morally bad. You are in danger of feeling righteous, and unwilling or unable to listen to people who take a different view. You are likely to seek the company of like-minded people, and to fear and despise the people who disagree with you. This is again ironic, as AI ethicists are generally (and rightly) keen on diversity.

The issues explored in the field of AI ethics are important, but it would help to clarify them if some of the heat was taken out of the discussion. It might help if instead of talking about AI ethics, we talked about beneficial AI, and AI safety. When an engineer designs a bridge, she does not finish the design and then consider how to stop it falling down. The ability to remain standing in all foreseeable circumstances is part of the design criteria, not a separate discipline called "bridge ethics". Likewise, if an AI system has deleterious effects it is simply a badly designed AI system.

PART TWO

AI TOMORROW

CHAPTER 8

A DAY IN THE LIFE

Julia woke up feeling rested and refreshed. This was unremarkable: she had done so ever since Hermione, her digital assistant, had been upgraded to monitor her sleep patterns and wake her at the best stage in her sleep cycle. But Julia could still remember what it was like to wake up to the sound of an alarm crashing through REM sleep and she felt grateful as she stretched her arms and smelled the coffee which Hermione had prepared.

Traffic feeds indicated that the roads were quiet, so with a few minutes in hand, Hermione updated her on her key health indicators – blood pressure, cholesterol levels, percentage body fat, insulin levels and the rest. Julia had long since stopped feeling disturbed by the idea of the tiny monitors that nestled in all parts of her body, including her bloodstream, her eye fluids, her internal organs, and her mouth.

She slipped into the outfit that she and Hermione

had put together last night with the aid of some virtual wardrobe browsing and a bit of online shopping. An airborne drone had dropped off the new dress while Julia was asleep; selecting the right size was rarely a problem now that virtual mannequins were supported by most retailers. The finishing touch was an intriguing necklace which Julia had her 3D printer produce overnight in a colour which complemented the dress perfectly, based on a design sent down by her sister in Edinburgh.

During the drive to the station, Julia read a couple of the items which Hermione had flagged for her in the morning's news feed, along with some gossip about a friend who had recently relocated to California. With a gesture inside the holosphere projected by Hermione for the purpose, Julia OK'd the micro-payments to the news feeds. Half an hour later, she smiled to herself as her self-driving car slotted itself perfectly into one of the tight spaces in the station car park. It was a while since she had attempted the manoeuvre herself; she knew she would not have executed it so smoothly even when she used to do her own driving. She certainly wouldn't be able to do it now that she was so out of practice.

The train arrived soon after she reached the platform (perfect timing by her car, again) and Hermione used the display in Julia's augmented reality (AR) contact lenses to highlight the carriage with the most empty seats, drawing on information from sensors inside the train. As Julia boarded the carriage the display highlighted the best seat to choose, based on her travelling preferences and the convenience of disembarkation at the other end of the journey.

Julia noticed that most of her fellow passengers wore opaque goggles: they were watching entertainments with fully immersive virtual reality (VR) sets. She didn't join them. She was going to be spending at least a couple of hours in full virtual reality during the meeting today and she had a personal rule to limit the numbers of hours spent in VR each day.

Instead she kept her AR lenses in and gazed out of the window. The train took her through parts of the English countryside where she could choose between overlays from several different historical periods. Today she chose the Victorian era, and enjoyed watching how the railway she was travelling was under construction in some parts, with gangs of labourers laying down tracks. She marvelled at that kind of work being done by humans rather than machines.

Hermione interrupted her reverie, reminding Julia that tomorrow was her mother's birthday. Hermione displayed a list of suggested presents for Julia to order for delivery later today, along with a list of the presents she had sent in recent years to make sure there was no tactless duplication. Julia chose the second gift on the list and authorised payment along with Hermione's suggested greeting.

That done, Hermione left her to indulge in her historical browsing for the rest of the journey. The augmented views became increasingly interesting as the train reached the city outskirts, and huge Victorian construction projects unfolded across south London.

Julia noticed that the content makers had improved the views since the last time she watched it, adding a host of new characters, and a great deal more information about the buildings in the accompanying virtual menus.

When she reached her office she had a half hour to spare before the meeting started, so she crafted an introductory message to an important potential new client. She used the latest psychological evaluation algorithm to analyse all the target's publicly available statements, including blog posts, emails, comments and tweets. After reading the resulting profile she uploaded it into Hermione to help with the drafting. Hermione suggested various phrases and constructions which helped Julia to keep the message formal, avoiding metaphors and any kind of emotive language. The profile suggested that the target liked all claims to be supported by evidence, and didn't mind receiving long messages as long as they were relevant and to the point.

It was time for the conference call – the main reason she had come into the office today. She was meeting several of her colleagues in VR because they were based in ten different locations around the world, and the topic was important and sensitive, so they wanted the communication to be as rich as possible, with all of them being able to see each other's facial expressions and body language in detail. Her VR rig at home wasn't sophisticated enough to participate in this sort of call.

A competitor had just launched a completely automated version of one of Julia's company's major service lines in two countries, and it would probably be rolled out worldwide within a couple of weeks. Julia and her colleagues had to decide whether to abandon the service line, or make the necessary investment to follow suit in automating it – which meant retraining a hundred members of staff, or letting them go.

As always, Julia was grateful to Hermione for discreetly reminding her of the personal details of the colleagues in the meeting that she knew least well. In the small talk at the start and the end of the call she was able to enquire about their partners and children by name. It didn't bother her at all that their ability to do the same was probably thanks at least in part to their own digital assistants.

Several of the participants in the call did not speak English as their first language, so their words were translated by a real-time machine translation system. The words Julia heard them speak did not exactly match the movement of their mouths, but the system achieved a very plausible representation of their vocal characteristics and their inflections.

A couple of times during the call Hermione advised Julia to slow down, or get to the point faster, using the same psychological evaluation software which had helped to craft the sales message earlier, and also using body language evaluation software.

After the meeting Julia had lunch with a friend who worked in a nearby office. Hermione advised her against the crème brûlée because it would take her beyond her target sugar intake for the day. Julia took a little guilty pleasure in ignoring the advice, and she was sure the dessert tasted better as a result. Hermione made no comment, but adjusted Julia's target for mild aerobic exercise for the day, and set a reminder to recommend a slightly longer time brushing and flossing before bed.

Before heading back to the office, Julia and her friend went shopping for shoes. She was about to buy a pair of killer heels when Hermione advised her that the manufacturer was on an ethical blacklist that Julia subscribed to.

Back in the office, Julia set about re-configuring the settings on one of the company's lead generation websites. The site used evolutionary algorithms which continuously tested the impact of tiny changes in language, colour and layout, adjusting the site every few seconds to optimise its performance according to the results. This was a never-ending process because the web itself was changing all the time, and a site which was perfectly optimised at one moment would be outmoded within minutes unless it was re-updated. The internet was a relentlessly Darwinian environment, where the survival of the fittest demanded constant vigilance.

Re-configuring the site's settings was a delicate affair because any small error would be compounded rapidly by the evolutionary algorithm, which could quickly lead to unfortunate consequences. She decided to do it with a sophisticated new type of software she had read about the day before. The software provider had supplied detailed instructions to follow with AR glasses. She went through the sign-up, login and operating sequences carefully, comparing the results with the images shown in her glasses.

By the time Julia was on the train home it was getting dark. She asked Hermione to review the recording of her day and download the new app her friend had recommended in passing at lunchtime. Then she queued up two short videos for the drive from the station to her house. She was the only person to get off the train at her home station, so she asked Hermione to display the camera feeds on the two routes she could take from the platform to her car. There was no-one on the bridge over the tracks but there was someone loitering in a shadowy part of the underpass. She took the bridge, even though it was a slightly longer route.

When she reached her car she checked that the delivery drone had opened her car's boot successfully and dropped the groceries there before re-locking it with the remote locking code. Happy in the full possession of her vegetables, she drove home, humming along to Joni Mitchell.

CHAPTER 9

INFORMATION INTO KNOWLEDGE

The science fiction writer William Gibson is reported as saying that "The future is already here – it's just not evenly distributed.[104]" Most of the things mentioned in the short story above are already available in prototype, and most of the rest is under development. It could take anywhere from five to thirty years for you to have working versions of all of them.

Some people will think the life described above is frightening, perhaps de-humanised. It is likely that more people will welcome the assistance, and of course generations to come will simply take it for granted. As Douglas Adams said, anything that is in the world when you're born is just a natural part of the way the world works, anything invented between when you're fifteen and thirty-five is new and exciting, and anything invented after you're thirty-five is against the natural order of things, and should be banned.[105]

Of course, there is no guarantee that the future will work out this way – in fact the details are bound to be different. For example we don't yet know whether the myriad devices connecting up to the Internet of Things will communicate with us directly, or via personal digital assistants like Hermione. Will you be reminded to take your pill in the morning because its bottle starts glowing, or will your version of Hermione alert you? No doubt the outcome will seem obvious in hindsight.

It has been said that all industries are now part of the information industry – or heading that way. Much of the cost of developing a modern car – and much of the quality of its performance - lies in the software that controls it.

Demis Hassabis, founder and CEO of DeepMind, has said that AI converts information into knowledge, which he sees as empowering people. The mission statement of Google, which owns his company, is to organise the world's information and make it universally accessible and useful. For many of us, most of the tasks that we perform each day can be broken down into four fundamental skills: looking, reading, writing, and integrating knowledge. AI is already helping with all these tasks in a wide range of situations, and its usefulness is spreading and deepening.

Marketers used to observe that much of the value of a product lies in the branding – the emotional associations which surround it. The same is now true of the information which surrounds it. You might think that the commercial success of a product as physical as say, skin cream does not rely on the provision of information to its consumers. Increasingly that is wrong. Consumers have access to staggering amounts of information about skincare, and many of them want to know how each product they might use would affect their overall regime. In a world of savvy consumers, the manufacturer which provides the most concise, easy-to-navigate advice is going to win market share.

From supermarket supply chains to consumer goods to construction to exploring for minerals and oil, the ability to crunch bigger and bigger data sets and make sense of them is improving pretty much every type of human endeavour. Kevin Kelly, the founder of Wired magazine, said the business plans of the next 10,000 startups are easy to predict: "Take X and add AI."[106] To coin a phrase, blessed are the geeks, for they shall inherit the Earth.

Andrew Ng, formerly of Google and Baidu, likes to say that AI is the new electricity. It is phenomenally powerful, increasingly ubiquitous, and it powers more and more of the things we rely on in our everyday lives. We will increasingly take it for granted, indeed ignore it, but if it was taken away we would howl with protest. It will also change everything. As Ng says: "Just as electricity transformed almost everything 100 years ago, today I actually have a hard time thinking of an industry that I don't think AI will transform in the next several years."[107]

CHAPTER 10

THE SHORT TERM: FRIENDS AND AUTOS

Voice control

In the coming years, we humans will increasingly communicate with machines by voice as well as by keyboard. This will make computers easier and faster to deal with, and will broaden the range of situations in which they are useful. It will also be helpful because many of the people who currently have no internet access and will be coming online in the next decade do not type. Typing is not a skill typically acquired by people living on a few dollars a day and unable to read.

Voice control is made possible by natural language processing, which we discussed in Part 1. At the moment this takes large amounts of processing power and considerable expertise, but the systems are getting more capable and more streamlined all the time. Two of the biggest applications are chatbots and digital assistants.

Chatbots enable organisations to provide useful information to their customers and stakeholders without the expense of employing lots of humans in call centres. They are a major focus of AI development work within businesses today. You may well have had an exchange with a machine on a company website already, thinking it was a human. This experience will become more common. Google unveiled a chatbot called Duplex in a stunning demo in May 2018. It proceeded to roll the system out slowly in pilot services, and at February 2020 it was available for limited applications (mostly restaurant bookings) in 48 US States[108], and a few other countries (mostly cinema bookings).[109]

In January 2020 Google announced the development of Meena, the world's new best chatbot. They measured its utterances against two tests: does it make sense, and is it specific. Humans achieve a score of 86%, and Meena scored 79%. The nearest competitor chatbot achieved only 56%.[110] And don't forget, we are still only at the beginning of our AI journey.

But the really big application of voice control and natural language processing will be our digital assistants. At the moment, Siri is a bit of a joke. But Siri is going to become serious.

Digital assistants

Siri, Alexa, Cortana, Google Assistant and the rest are primitive today, but in a decade or two their descendants will be our constant companions, and we will wonder how we ever got along without them. They will be our gateway to the internet, and our invaluable assistants as we navigate our way through the world. Among other things, they will negotiate with and filter out most of the Internet of Things. Although we may not notice it, this will be a blessed relief. Imagine living in a world where every AI-enabled device has direct access to you, with every chair and handrail pitching their virtues to you, and every shop screaming at you to buy something. This dystopia was captured in the famous shopping mall scene in the 2002 film "Minority Report", and more laconically in Douglas Adam's peerless "Hitchhiker's Guide to the Galaxy" series, where the Corporation that produces the eponymous guide has installed talking lifts, known as happy vertical people transporters. They are extremely irritating.

Friends

What generic name will be adopted for these assistants? Most of the essential tools which we use every day have one-syllable names, like phone, car, boat, bike, plane, chair, stove, fridge, bed, gun[111]. Those which have two syllables are often elided or rhyming, like iron and hi-fi. A few, like hoover, are named after the person or company who made the first successful version.

As yet we have no short name for our digital assistants. "Digital personal assistant" and "virtual personal assistant" both capture the meaning but are hopelessly unwieldy. Maybe we'll initialise them, like TVs, and call them DAs, DPAs, or VPAs. Or maybe we'll use the brand name of one of the early leaders, and call them all Siris. Google's chairman Eric Schmidt came up with the interesting idea that we'll find ways to name them after ourselves, and his would be called "not-Eric".[112] Perhaps we'll channel Philip Pullman's His Dark Materials trilogy and call them daemons. Or perhaps - and this is my favourite - we'll call them our "Friends".

Not by voice alone

We will probably also need to invent a new type of interface to enable us to communicate with our digital assistants. The 2013 movie "Her" is one of Hollywood's most intelligent treatments of advanced artificial intelligence. (I realise that isn't saying much, but for better or worse, Hollywood does give us many of the metaphors that we use to think about and discuss future technologies.) The essence of the plot is that the hero falls in love with his digital assistant, with intriguing consequences. Although he uses keyboards occasionally, most of the time they communicate verbally.

There will be times when we want to communicate with our "friends" silently. Portable "qwerty" keyboards will not suffice, and virtual hologram keyboards may take too long to arrive – and they may feel too weird to use even if and when they do arrive. Communication via brain-computer interfaces will take still longer to become feasible, so perhaps we will all have to learn a new interface – maybe a one-handed device looking something like an ocarina[113].

Another possibility is sensors – perhaps embedded as tattoos - on the face and around the throat which have micro-sensors to detect and interpret the tiny movements when people sub-vocalise, i.e., speak without actually making a noise. Motorola, now owned by Google, has applied for a patent on this idea.[114] Perhaps when we are out and about in the future, we will get used to seeing other people silently miming speech as we all chat happily to our "Friends".

Choosing our Friends

Digital assistants will be very big business, and the evolution of their industry will be fascinating. Will it turn out to be a natural monopoly, where the winner takes all? If so, the winner will find itself the subject of intense regulatory scrutiny, and probably of moves to break it up or take it into public ownership. Or will there be a small number of immensely powerful contenders, as in the smartphone platform business, where Apple and Android have the field almost to themselves?

Will we all choose one brand of Friend at an early age, or during adolescence, and stick with it for life, as many people do with smartphones? Doubtless the platform providers will seek to lock us in to that kind of loyal behaviour. Or will we be promiscuous, hopping from one provider to the next as they jostle and elbow each other, taking turns to launch the latest, most sophisticated software?

Wearables, insideables

At the moment, the device which hosts the primitive forebears of these essential guides is the smartphone, but that is merely a temporary embodiment. We will surely progress from portables to wearables (Apple

Watch, Google Glass, smart contact lenses...) and eventually to "insideables": chips inside our bodies.

You doubt that Google Glass will make a comeback? The value of a head-up display, where the information you want is displayed in your normal field of vision, is enormous; that's why the US military is happy to pay half a million dollars for each head-up display helmet used by its fighter aircraft pilots.

Apple Watch has been successful because people will pay good money to simply raise their wrist rather than go to all the bother of pulling their smartphone out of their pocket. How much better to have that hunger for the latest bit of gossip sated, and that essential flow of information about your environment displayed right in front of your eyes with no effort whatsoever?

Head-up displays will get better and better until they finally yield to "insideables". The technology to enable a chip implanted inside you to insert a detailed image directly into your brain is far ahead of where we are now. But it is the next logical step in the process after wearables. It is just a question of how long.

Screens will be everywhere by this time, of course: on tables, walls both interior and exterior, on the backs of lorries so that you can see what is ahead of them.[115] But we will probably want to carry our own screens around with us, not least because we won't always want other people to see what we are looking at.

Self-driving vehicles: why introduce them?

As we saw in chapter 6, the current project to develop self-driving vehicles was kick-started by DARPA in 2004. The case for developing them is simple and overwhelming: around the world, human drivers kill 1.35 million people a year. Road traffic accidents are the leading cause of death for people aged 5 to 29[116].

90% of these accidents are caused by human error.[117] Humans become tired, angry, drunk, sick, distracted or just plain inattentive. Machines don't, so they don't cause accidents. To paraphrase Agent Smith in "The Matrix", we are sending humans to do a machine's job.

There is also the wasted time and frustration. We all know that driving can be fun, but not when you're stuck in traffic – perhaps because one of your fellow humans has caused an accident. On average, American commuters spend the equivalent of a full working week stuck in traffic every year – twice that much if they are lucky enough to work in San Francisco or Los Angeles.[118] We drive rather than use public transport because there is no appropriate public transport available, or sometimes because we prefer travelling in our own space. Self-driving cars could give us the best of both worlds, allowing us to read, sleep, watch video or chat as we travel.

Finally, self-driving cars will enable us to use our environments more sensibly, especially our cities. Most cars spend 95% of their time parked.[119] This is a waste of an expensive asset, and a waste of the land they occupy while sitting idle. We will consider later how far self-driving cars could alleviate this problem.

To autonomy and beyond

Self-driving cars, like our artificially intelligent digital assistants, are still waiting to receive their generic name. "Self-driving cars" is the name we are stuck with for the time being, but it is all clunk and no click[120]. At the end of the 19th century it was becoming obvious that horseless carriages were here to stay, and needed a shorter name. The Times newspaper adopted "autocar" but the Electrical Engineer magazine objected that it muddled Greek (auto) with Latin (car). It argued instead for the etymologically purer "motor-car".[121] Perhaps we will contract the phrase "autonomous vehicle", and call them "AVs", or "Autos".

Some people are going to hate self-driving cars, whatever they are called: petrol-heads like Jeremy Clarkson are unlikely to be enthusiastic about the objects of their devotion being replaced by machines with all the romance of a horizontal elevator. Some people are already describing a person who has been relegated from driver to chaperone as a "meat puppet".[122]

The levels of autonomy were defined in January 2014 by the Society of Automotive Engineers (SAE) a US-based, but global standards organisation:

Level 0: No automation

Level 1: Driver assistance. The vehicle can handle either steering, or throttle and brakes, but the driver must always be prepared to take over.

Level 2: Partial assistance. As level 1, but the vehicle can handle both steering, and throttle and brakes.

Level 3: Conditional assistance. The vehicle monitors its surroundings and in normal circumstances takes care of all steering, throttle and brakes. But it cannot handle all contingencies, and the driver must always be prepared to take over.

Level 4: High automation. The vehicle handles all driving within a pre-determined geography and / or road conditions.

Level 5: Full automation. The driver sets the destination and starts the vehicle and the vehicle does the rest – in all circumstances.

Google's initial idea was that the first self-driving cars in general use would be Level 3, but discovered that once the test drivers considered the technology to be reliable, they became complacent and engaged in "silly behaviour". For instance, one turned round to look for a laptop in the back seat when the car was doing 65 mph. This experience persuaded Google to hold back the release of their cars until they reached Level 4.[123]

The count-down

Engineers joke that the first 99% of a large project takes the first 99% of the time allocated, and the last 1% takes the second 99%. We are now in that last 1% of the Autos project, and no-one knows for sure how long it will last.

Waymo and others are proceeding with great caution, and this is right and proper. The death toll from human driving is appalling, but we humans are far more scared of being killed by a robot than by another human. In an entirely rational world, Autos would be introduced as soon as they are even slightly safer than human drivers. In our human, emotional world, they cannot be introduced until they are substantially better. It is interesting how little response there was from the broad public when an Uber car became the first Auto to kill a human, but it would not take many such incidents to prompt a backlash against the technology, which would delay its introduction significantly.

There is no consensus about when Autos will be ready for prime time, and when they will become a common sight on our streets. Some developers and informed observers think it will be soon, while others think it is many years away. Some of this disagreement is a squabble over terminology. Level 5 Autos are probably quite a few years away, but Level 4 is all you need to operate a commercially compelling taxi or delivery service. Given that Google's Waymo is already driving members of the public who have not signed NDAs around in cars with nobody in the front seats, it seems unlikely that Level 4 cars for commercial services like taxis are far off.

Another interesting data point is that in March 2020, Waymo raised $2.25bn of external equity from some of the smartest investors in Silicon Valley, who obviously believe that the technology will be ready for prime time within years rather than decades.[124]

Overall, it seems reasonable to expect that self-driving technology will be ready for prime time by the middle of the 2020s.

The impact of Autos on cities

Enthusiasts for self-driving cars sometimes paint a

utopian picture of cities where almost no-one owns a car because communally-owned taxis are patrolling the streets intelligently, anticipating our requirements and responding immediately to our summons. Today, our cars sit idle 95% of the time, squatting like polluting toads on vast acres of city land. In this bright tomorrow they are used efficiently, and the land given over to parking can be returned to pedestrians and useful buildings. Traffic flows smoothly because the cars are in constant communication with each other: they don't bunch into jerky waves and they don't need to stop at intersections.

This is almost certainly an exaggeration. There will still be peak times for journeys, so even if most journeys are undertaken in communal cars, many of them will be parked during off-peak hours. (One dystopian scenario has hordes of "zombie" cars driving round cities in circles, prowling for passengers and never stopping, in order to avoid incurring parking fees.) Traffic will still have to halt at intersections every now and then if pedestrians are ever going to be able to cross the road. Not every pedestrian crossing can have a bridge or an underpass.

Nevertheless, machine-driven cars will be more efficient consumers of road space than human drivers. Traffic conditions are not fixed fates which once imposed can never improve. A congestion charge has significantly reduced traffic flows in London, and the switch to almost-silent hybrid taxis has made walking the streets of Manhattan a less tiring experience than it used to be.[125] In any case, more efficient road use is not required to justify the introduction of self-driving cars. The horrendous death and injury toll imposed by human drivers is sufficient, together with the liberation from the boredom and the waste of time caused by commuting.

The ethics of self-driving cars

Imagine that your self-driving car is travelling down the road, minding its own business, when a child, unpredictably, dashes across the street ahead of you. Calculating at super-human speed, it analyses the only three available options: brake hard and then either maintain direction, or swerve right, or swerve left. Even though it has already applied the brakes far quicker than any human could have, it calculates that these options will result in the death, respectively, of the child, of an adult bystander, or of you, its passenger. Which option should it select?

With grim humour, some have suggested that the answer will vary by car. Perhaps a Rolls Royce will always choose to preserve its owner, while a cheaper model may accord its occupants less respect.

In theory, what is happening here is the extension of human control over the world: the arrival of choice. Today, 27% of the victims of traffic accidents are pedestrians and cyclists. What happens to them and the drivers of the cars which hit them is currently decided by the skill of the driver and blind chance. In future we will have more power to affect outcomes, and with increased power comes increased responsibility.

However, in reality, self-driving cars will rarely if

ever be required to make decisions like this. First, because they will not place themselves in situations where such choices are needed. They will observe speed limits, and their thousands of sensors will continually be testing every aspect of the vehicle's safety equipment. They will be the safest vehicles our roads have ever seen. And second, because swerving would raise the likelihood of myriad unforeseen consequences, and would generally be too risky to attempt. Self-driving car experts do not believe this sort of consideration is slowing down the development process, nor do they think it should.[126]

In chapter 18 we will consider the impact of self-driving cars on employment.

CHAPTER 11
RELATED TECHNOLOGIES

One ring to bind them

Artificial intelligence is increasingly our most powerful technology, and it will increasingly inform and shape everything we do. Its full-blooded arrival coincides with the take-off of a series of other technologies. They are often driven at least in part by AI, and they will all impact the way our societies evolve.

Because they will all unfold in different ways and at different speeds, it is impossible to predict exactly what the individual and combined impact of these interlacing technologies will be, other than that it will be profound.

The Internet of Things

The Internet of Things (IoT) has been talked about for years – the term was coined by British entrepreneur Kevin Ashby back in 1999.[127] Indeed it has been around for long enough to have acquired a selection of synonyms. GE calls it the Industrial Internet, Cisco calls it the Internet of Everything, and IBM calls it Smarter Planet. The German government calls it Industrie 4.0[128], the other three being the introduction of steam, electricity, and digital technology.

My favourite alternative name for the IoT is Ambient Intelligence,[129] which comes nearest to capturing the essence of the idea, which is that so many sensors, chips and transmitters are embedded in objects around us that our environment becomes intelligent – or at least, intelligible.

When originally conceived, the IoT was based on Radio Frequency Identification tags (RFID), tiny devices about the size of a grain of rice which can be "read" remotely without being visible to the device which "reads" them. The RFID is a passive device, and this concept does not involve any AI.

Later, technologies like Near Field Communication (NFC) were developed, which allow for two-way data exchange. Android phones have been NFC-enabled since 2011, and it powers the Apple Pay system which was launched with the iPhone 6.

The IoT is becoming possible because the component parts (sensors, chips, transmitters, batteries) are becoming cheaper and smaller at – you guessed it – an exponential rate. At the end of 2019 there were around 5 billion IoT-connected devices,[130] and market intelligence firm IDC forecasts 40 billion of them by 2025.[131] Many of these devices have multiple sensors – smartphones can have as many as 30 each.[132]

Looking further ahead, the internet entrepreneur Marc Andreessen predicts that by 2035, every physical item will have a chip implanted in it. "The end state is fairly obvious – every light, every doorknob will be connected to the internet."[133]

Making the environment intelligible offers tremendous opportunities. A bridge, building, plane, car or refrigerator with embedded sensors can let you know when a key component is about to fail, enabling it to be replaced safely without the loss of convenience, money, or life which unforeseen failure might have caused. This is known as condition-based maintenance, or predictive maintenance, and is being pioneered with encouraging results by, for instance, MTR Corporation, which runs Hong Kong's urban transit network.[134]

The IoT will also improve energy efficiency across the economy, as the heating or cooling of buildings and vehicles can be regulated according to their precise temperature, humidity, etc., and the number and needs of the people and equipment using them.

Since its launch in 1990,[135] the world-wide web has rendered our lives immeasurably easier, by placing information at our fingertips. The IoT will take that process an important stage further, by dramatically improving the amount and quality of information, and enabling us to control many aspects of our environment. You will be able to find out instantaneously the location and price of any item you want to buy. You will know the whereabouts and welfare of all your friends and family – assuming they don't mind – and the location of all your property: no more lost keys! You will be able to control at a distance the temperature, the volume, the location of things that you own. Your own health indicators can be made available to anyone you choose, which will certainly save many lives.

Many of the applications made possible by the IoT will be surprising. For instance, perhaps it will transform our approach to the punishment of crime. It is obvious to most people that our current approach is not working. One in every 1,000 adults in the US is currently in jail, and recidivism rates show that all this achieves is to condemn many of them to a life (often short) of crime and drug addiction. Supporters of the status quo argue that the system keeps the public safe by removing criminals from general circulation, and it provides a measure of justice, but few would argue that it is good at reforming behaviour or rehabilitation. Reformers argue that removing the liberty of criminals is sufficient punishment, and it would be better to do that without also institutionalising them and driving them back into crime when released. They believe that restorative justice, in which criminals do as much as possible to undo the harm they have done, is better all round than purely retributive justice.

IoT technologies like wearable devices which cannot be removed, coupled with sophisticated sensors in the immediate environment, could be deployed to restrict the freedom of convicted criminals, while allowing them to live in humane conditions. This would make it less likely that they will re-offend after their term is complete. The idea is controversial, with some people complaining that prisoners will have "cushy" sentences, and others fearing the arrival of Big Brother. But the technology at least allows discussion of alternatives to a system that is currently broken.[136]

Like any powerful technology, the IoT will raise concerns, particularly about privacy and security. It will also need a set of standards, so that all those semi-intelligent chairs and cars talk the same language. This may come about through government regulation, industry co-operation, or because one player becomes strong enough to impose its standards on everyone else.

Robots

The idea of mobile, artificial devices which carry out complicated manoeuvres, usually to save human labour, is an old one, going back to at least the ancient Greeks. The word "robot" was first introduced in 1921 by Karel Čapek in his play "R.U.R." (Rossum's Universal Robots), and since then a great deal of time and money has been spent trying to develop them, but we are still a very long way from having anything like Rosie the household robot from the 1960s TV series "The Jetsons".

Robots are being developed for a myriad of applications, including heavy industry, healthcare, hospitality, domestic service, military, hazardous environments and so on. As with AI, we are still in the early days, and the great majority of robots in actual use are specialised and expensive machines deployed in factories. In 2018, the latest year for which data is available from the IFR (International Federation of Robotics), 380,000 industrial robots were shipped. More than a third of these,

132,000, were shipped to China, the biggest market. China had 97 robots per 10,000 manufacturing workers, comparable to Europe's 106, and the USA's 91. It was significantly behind leader, South Korea, at 710.[137]

Machine learning, and in particular reinforcement learning, is being used to help robots like BRETT (Berkeley Robot for the Elimination of Tedious Tasks) acquire skills, but it is a slow process.[138] One of the tasks BRETT is working on is folding towels. A Japanese company called Seven Dreamers was seen as the leader in this field, but in 2019 it filed for bankruptcy as investors lost patience.[139]

Researchers are trying out different ways to improve robot performance. Teams at Carnegie Mellon University in Pittsburgh and at Google are getting robots to learn about their physical environment by having them simply prod, poke, grasp and push objects around on a table-top, in much the same way that a human child learns about the physical world. Having collected a large data set from this activity, the systems turn out to be better at recognising images from the ImageNet database than systems which have not had the physical training.[140]

Google's robot army decamps to Softbank

In late 2013, Google announced the purchase of no fewer than eight robotics companies. (Since you ask, they are Boston Dynamics - purveyor of the famous Big Dog and Atlas models - Bot and Dolly, Meka, Holomni, SCHAFT, Redwood, Industrial Perception, and Autofuss.) Google also announced that the new division which owned them would be run by Andy Rubin, who had created a huge global business with the Android phone platform.

A year later, in October 2014, Andy Rubin left Google to found a technology startup incubator, and it became apparent that Google had not managed to integrate its robotic acquisitions satisfactorily. In particular, the military focus of Boston Dynamics sat uneasily with Google's culture. In June 2017 it was announced that the Japanese investment group Softbank had acquired both Boston Dynamics and Schaft. A Japanese technology firm, Softbank is Japan's third-largest public company. It had already acquired another leading robotics company, France's Aldebaran, maker of the cute-looking and child-sized Pepper robots, in 2012.[141]

Softbank's founder Masayoshi Son, believes that a singularity will happen by 2047. In January 2017 he announced the formation of Vision, an investment fund with $100 billion to invest in high-technology businesses in preparation for this event.[142] The immediate future of the robot army looked less rosy when Softbank ran into trouble in 2019 and early 2020 with a number of its largest investments, including workspace company WeWork.

Robots in the warehouse

Kiva Systems was established in 2003, and acquired by Amazon in 2012. Kiva produces robots which collect goods on pallets from designated warehouse shelves and deliver them to human packers in the bay area of the warehouse. Amazon paid $775m for the nine-year old company and promptly dispensed with the services of its sales team. Renamed Amazon Robotics in August 2015, it is dedicated to supplying warehouse automation systems to Amazon, which obviously considers them an important competitive advantage. As at June 2019, Amazon had 200,000 Kiva robots in its warehouses.[143]

Complicated relationships

It is going to take us humans a while to get used to having robots around. Pepper is 120 cm tall and costs around $1,200. It has a limited ability to "read" human emotions and respond appropriately. It has proved extremely popular in Japan, with four batches of 1,000 selling out in less than a minute when they went on sale in September 2015. The response to Pepper has not been straightforward, however. The manufacturer felt obliged to outlaw any attempt to engage in sex with the robot, and a Japanese man was prosecuted for assaulting one when drunk.[144]

A robot called Hitchbot managed to cross Canada from coast to coast in 2015, but was attacked and decapitated in Philadelphia when it tried to repeat the performance in the US.[145]

More robots: androids and exoskeletons

It is not clear that robots need to resemble humans closely to perform their tasks, but that doesn't stop researchers from trying to make them. (Robots with human appearance used to be what the word "android" meant before Google appropriated it for phone software.) We are probably quite a few years away from having robots with the verisimilitude of the ones in the film "Ex Machina", or the TV series "Humans", for example. Nadine is a state-of-the-art prototype working as a receptionist at Nanyang Technological University in Singapore. It is humanoid, but doesn't fool anyone who takes a second glance.[146] Modelled on its inventor, professor Nadine Thalmann, it cannot walk, but it can smile, turn its head, and shake your hand. Its voice is powered by an AI similar to Siri.

In 1970, a Japanese professor of engineering named Masahiro Mori coined a phrase which was later translated as "uncanny valley".[147] This describes how a robot's level of appeal to humans increases as it becomes more humanoid, then dips for a while, before returning to high appeal as it approaches complete verisimilitude.

Many robots will be special-purpose devices, constructed to carry out a very specific task. An example is the Grillbot, a robot the size of a table tennis bat which cleans your barbecue grill, and is otherwise entirely useless.[148]

Some people argue that exoskeletons are wearable robots. Whether or not that is semantically correct, they will certainly enable one human to do the work of several. At the moment, leading companies in the space like Ekso Bionics[149] are focusing on patient rehabilitation systems. But before long similar equipment will be available for people carrying out physically demanding tasks in the military, manufacturing, and distribution.

Drones

Another form of robot which is taking off fast is unmanned aerial vehicles (UAVs), or drones – flying machines that can be autonomous, or controlled remotely. They have a wide range of potential applications, including taking surreptitious photos of celebrities, taking selfie videos for life-logging Millennials, delivering parcels for Amazon, monitoring buildings and crops, and killing people. Internet-connected drones with powerful sensors and computers on board are quickly becoming essential tools for companies in the utilities and engineering industries, as well as government agencies.[150]

In Switzerland, a drone company called Matternet received approval in 2017 to start flying medical supplies to hospitals in urban areas aboard robotic quad-copters.[151] And the Icelandic Transport Authority gave a company called Flytrex approval to fly delivery drones out of the line of sight. Initially, customers would have to pick up deliveries from a designated drop-off zone, but the company expected approval shortly to fly direct to their backyards.[152]

The first company to take delivery drones mainstream may well be Amazon. Its CEO Jeff Bezos first announced the formation of a drone business unit called Amazon Prime Air in December 2013. The plan is to have drones weighing 25kg, carrying items weighing 2.3kg, at speeds up to 50 mph, between 200 and 500 feet up, up to 10 miles from a warehouse.[153] Getting regulatory approval has been challenging, and although a launch was said to be imminent in June 2019, it had not happened by spring 2020.[154]

The serious challenges that drones pose for regulators was graphically illustrated in December 2018, when malicious drone operators forced the closure of Gatwick airport, Britain's second-largest. Tens of thousands of people had their Christmas holidays disrupted,[155] and the inability of the authorities to anticipate and prevent the disruption, or halt it for several days after it started, or even catch the perpetrators afterwards, was a major embarrassment for the country. The government never explained its failure to invest in the capability to protect one of Europe's ten busiest airports against such a blindingly obvious threat.[156]

Probably the best-known application of drones to date is military. Drones have played a major role in the USA's global response to the attacks on 11 September 2001, being deployed in Afghanistan, Pakistan, Yemen, Somalia, and Libya. The first known fatal US drone strike occurred in Pakistan in 2004, and killed six people, including two children.[157]

In 2014 it was reported that the US operated over 10,000 military drones, and was training more pilots to operate drones than to fly bombers and fighters combined.[158]

In March 2019, President Trump signed an executive order relieving the US intelligence community of the obligation to report its fatal drone attacks, but the US Defense Department continues to publish an annual report on fatal drone attacks.[159] Accurate statistics are inevitably very hard to come by. The US think tank

The New America Foundation reckons that from 2004 to 2011, between 1,400 and 2,100 people were killed by US drones in Pakistan alone.[160]

Virtual and Augmented Reality: developments

In 2014, many people got their first taste of virtual reality (VR) from Google Cardboard, an ingenious way to let smartphones introduce us to this extraordinary technology. Users download compatible VR apps to their phone, then place the phone into a cheap viewer and view the content through the viewer's lenses.[161] By March 2017, more than 160 million apps had been downloaded, and by November 2019, 15 million viewers had been sold or given away.[162] Google followed Cardboard with an enhanced unit called Daydream, but it did not prove popular enough, and was discontinued in October 2019.[163]

2016 was widely expected to be the year that VR really took off, as Facebook's Oculus VR launched Rift, the first VR equipment for consumers that offered high definition visuals and no latency.

(Latency is a failure of synchronisation between the stimuli from different sources reaching the brain: if your visual experience is out of synch with your other senses, your brain gets confused and unhappy, and can make you feel sick.[164] When VR is effective, it is surprisingly powerful. When the sense data being received by the brain become sufficiently realistic, the brain "flips", and decides that the illusion being presented is the reality.)

In the event, Oculus Rift was not the breakthrough

product that many expected. Consumers found the equipment expensive, and heavy to wear. VR rivals Sony and HTC fared better, but overall, the installed base remained small in early 2017, at less than 0.5% of the number of accounts on Steam, the world's biggest games distribution platform.[165]

In 2016-17, AR stole the limelight from VR when a developer called Niantic achieved startling success with an AR version of Pokemon Go. Augmented reality (AR) is different to VR in that it is overlaid on your perception of the real world rather than replacing it. It can make elephants swim through the air in front of you, or plant a skyscraper in your back garden. This is handy if you want to remain alert to the threat from cars and potholes while you are out and about, hallucinating swimming elephants.

A week after the launch of Pokemon Go in July 2016, 28 million people were using it every day, looking for virtual Pokemon characters to capture. The app was downloaded 650 million times by February 2017[166], but the enthusiasm could not be sustained, and by September it was no longer among the iPhone's top 200 apps.[167]

Microsoft attempted to mark out a distinct territory for its own AR offering, the Hololens, by called it mixed reality. In mixed reality (which is not usually abbreviated to MR), the virtual object is "anchored" in the real world so that you can, for instance, walk around it and manipulate it. In many forms of AR, you only see the virtual object when you look at the screen of your smartphone. Hololens requires a headset, so the object is always present in your field of view. A startup called Magic Leap raised $1.3bn for this type of AR from investors including Google, with a series of impressively realistic demos.[168]

Apple launched AR development software called ARKit in June 2017. The software that enables consumers to use the AR was unveiled in a surprisingly downbeat fashion in September 2017, perhaps to avoid the backlash of disappointment which has hit previous incarnations of the technology. But with an installed based of hardware which means millions of potential immediate users, Apple is well placed to play an important role in bringing AR into the mainstream. There are persistent reports that Apple will launch augmented reality glasses in 2021 or 2022, called simply Apple Glass.[169]

Google's first foray into augmented reality glasses, Google Glass, caused a stir, but ultimately failed as a consumer product. But it has not given up, and in June 2020 it bought Canadian smartglass manufacturer North.[170]

Facebook still sees VR as the next major platform after smartphones, and in late 2019 it announced the launch of Facebook Horizon, a virtual world where people can meet, talk, and play games together. (An earlier virtual world called Second Life, launched by Linden Labs in 2003, grew to a million users in 2013. It is still operating, although the monthly under number had halved by 2018.[171]) Early previews of Facebook Horizon said the avatars and the overall experience seemed more like a Pixar / Disney movie than the typically darker and more sophisticated world of modern video games.[172]

Applications of VR and AR

Insofar as there is any debate about whether VR is going to be an important development, it's between those who think it's going to be huge, and those who think it's going to change everything. Research company SuperData reported global VR revenues (hardware and software) at $3.6bn in 2018, 30% up from 2017, and forecast it to exceed $16bn in 2022. AR revenues were $2.3bn in 2018 and forecast for $17.8bn in 2022.[173]

The biggest application in the short term is expected to be video games, which is a huge market: gaming has for some time rivalled Hollywood for leadership in global sales of packaged entertainment.[174] Judging by the content that was made available for Google Cardboard and Daydream, people also enjoy ersatz travel, and adventurous experiences. VR versions of Google Street View let you wander around Manhattan until the latency makes you ill, and other developers offer you rollercoaster rides, and adventure sports from skiing to hang gliding.

In the longer term the potential applications are bewildering. Without ever leaving our armchairs we should be able to enjoy such realistic simulations of events like sports matches and music concerts that we will question the value of struggling with transport and crowds to attend the real thing. Of course, the crowd has a lot to do with making the event exciting in the first place, so the organisers of VR events will want to find a way to recreate the effect of being in a crowd. Except that you'll be able to sit next to your friend, who happens to be in a VR rig a couple of continents away at the time.

Education and informal learning is also likely to experience a VR revolution. How much more compelling will it be to learn about Napoleon by experiencing the battle of Waterloo than by reading about it, or listening to a lecturer describe it? How much easier will it be for a teacher to explain the molecular structure of alcohol by escorting her pupils round a VR model of it?

Businesses will find many uses for VR, and because they often have larger budgets than consumers and educational institutions, they may sponsor the creation of the most cutting-edge applications. Computer-aided design environments will become startling places to work, for instance, allowing architects, designers and clients to explore and discuss buildings in great detail before ground is broken. And who knows what uses the military will find for VR. One frightening thought is that VR could become a powerful and truly terrifying instrument of torture.[175]

Telecommunication will also be taken to a new level. Although audio-only phone calls still predominate, good video-conferencing facilities add enormously to the effectiveness of a long-distance conversation, and the additional step of feeling present in the same space will improve the experience again. Anything which involves your re-location in time or space should be fertile ground for VR.

The applications described above all seem extremely

likely to happen at some point. The question (to which no-one currently has the answer) is when. On the other hand, it is not yet clear whether VR will turn out to be a good medium for movies. In a film, the director wants to direct your attention, and it isn't helpful if half the audience is busy gawping at images or events 180 degrees removed from the focus of the action.

Cynics will point out that new media (TV, video, the internet were all new media in their early days) are established only when users have found ways to apply them to porn, gambling, and then sport. No doubt VR will make its contribution to these areas of human activity, but I'm not going to get sucked into a discussion of what could be achieved with haptic suits – clothing which allows users to experience sensations of heat and touch initiated remotely by someone else.

The death of geography

The death of geography has been declared numerous times, but despite the rise of telephony, digitisation and globalisation, business and leisure travel just keeps on growing. Could the sense of genuine "presence" which good VR confers finally make the old chestnut come true? Will talent continue to be drawn into the world's major cities, or will VR puncture their inflated real estate prices, and smear humanity more evenly across the planet?

Maybe virtual reality can render scarcity less valuable, and less problematic. In the real world, not everyone can live in a beautiful house on a palm-fringed beach, drive an Aston Martin, and be greeted by a Vermeer as they enter their living room. With virtual reality, everyone can – to a fair degree of verisimilitude.

The world-wide web has given us something like omniscience, and virtual reality looks set to give us something like omnipresence. Perhaps all we will need after that is a technology to give us something like omnipotence.

CHAPTER 12

DISRUPTION

The exponential improvement of computing and AI will bring enormous benefits, but this means change, and change is uncomfortable.

Business buzzwords

In the early 2010s the hottest buzzword (buzz-phrase?) in business circles was Big Data. As always, there was a good reason for this, even if the endless repetition of the phrase became tiresome. Executives woke up to the fact that the huge amount of information they had about their customers and their targets could finally be analysed and turned into useful insights thanks to the reduced cost of very powerful (parallel processing) computers and clever algorithms. At long last, consumer goods manufacturers could do better than simply grouping their customers into zip codes, or broad socio-demographic tribes. They could find tiny sub-clusters of people with identical requirements for product variants and delivery routes, and communicate with them in a timely and accurate fashion.

Previous disruption by technological innovation

In the mid-2010s the buzzword became digital disruption, and again there is good reason for it. But the disruption of businesses and whole industries by technological innovation is nothing new.

On 13th October 1908, a German chemist named Fritz Haber filed a patent for ammonia, having managed to solidify nitrogen in a useful and stable form for the first time: three atoms of hydrogen and one of nitrogen. Nitrogen is a basic nutrient for plants, and it makes up 78% of the atmosphere, but its gaseous form is hard for farmers to use.

While Haber was developing his technique, thousands of men were labouring in the hot sun of Chile's Atacama desert, digging saltpetre (nitrate) out of the ground. Apart from guano from Peru, Chile's saltpetre was the world's only source of nitrogen in solid form. As soon as Haber announced his discovery, saltpetre became uneconomic, and if you visit the Atacama today you can still see substantial ghost towns which were abandoned soon after October 1908.

Today's disruption is caused by the digital revolution – the internet, to be specific – and unlike previous examples of disruption, it is affecting many industries simultaneously. During the dotcom boom and bust at the turn of the century there was much talk of companies being "disintermediated" by the internet, and a continuing example of that is the publishing industry, where publishers no longer dictate whose books get read because Amazon has enabled authors to publish themselves.

Kodak

The poster child of digital disruption is Kodak. At its peak in the late 1980s it employed 145,000 people and had annual sales of $19bn. Its 1,300-acre campus in Rochester, New York had 200 buildings, and George Eastman was as revered there a few decades ago as Steve Jobs is in Silicon Valley today.

Kodak's researchers invented digital photography, but its executives could not see a way to make the technology commercially viable without cannibalising their immensely lucrative consumer film business. So other companies stepped in to market digital cameras: film sales started to fall 20-30% a year in the early 2000s, even before the arrival of smartphones. Kodak is often accused of being complacent, but the dilemma it faced was almost impossible. In the classic phrase of the dotcom era, it needed to eat its own babies, and this it could not do.

Instead it spent a fortune entering the pharmaceuticals industry, paying $5.1 billion for Sterling Drug, and another fortune entering the home printing industry. Both ventures were unsuccessful, and Kodak filed for bankruptcy in 2012, emerging from it a year later as a mere shadow of its former glory. Today it has annual sales of $2bn and employs 8,000 people. 80 of the 200 campus buildings have been demolished and 59 others have been sold off. Its market capitalisation at the time of writing is about $92 million, down from $1,500 million in 2005.[176]

Peer-to-peer

One of the most disruptive developments so far this century has been the rise of peer to peer business models, in which two consumers transact business directly, using a platform provided by a commercial third party. The leading practitioners are Airbnb and Uber, both founded in San Francisco, of course – in 2008 and 2009 respectively.

The level of investor enthusiasm for the peer-to-peer model is demonstrated by comparing Airbnb's valuation of $38bn in May 2018[177] is four times higher than Hyatt's $9bn.[178] Yet Hyatt has over 913 hotels around the world,[179] 127,000 employees,[180] and revenues in 2019 of $5bn.[181] Airbnb owns no hotels, has 5,600 employees,[182] and in 2017 it made a profit of $93m on revenues of $2.6bn.[183]

Prior to the virus crisis, Airbnb was planning to float on the stock market via an Initial Public Offering (IPO) during 2020, following in the footsteps of Uber, which did so in May 2019. Even without the virus, this was going to be tricky, since Uber's share price dropped sharply after the IPO.

This sort of growth is unsettling for incumbents. Taxi drivers around the world protest that Uber is putting them out of business by competing unfairly. Hoteliers have tried to have Airbnb banned from the cities where they operate, sometimes successfully. The peer to peer giants are often portrayed as examples of a "move fast and break things" culture endemic to Silicon Valley, named after an unofficial motto which Facebook abandoned in 2014, in which selfish and ignorant white young men cause untold damage to culture and infrastructure.

Some of the criticism is justified. Uber's founder and CEO, Travis Kalanick, was aggressive and confrontational, and riled many city authorities. In June 2017 he resigned amid allegations that he tolerated and even promoted an unhealthy corporate culture. The proliferation of Airbnbs in tourist hotspots like Barcelona and New York have reduced the supply of rental properties for residents of those locations, and some of them have become party locations, and a serious nuisance for their neighbours.

Some of the criticism is questionable, or downright spurious. In some cities, taxi licenses make huge profits for their owners – often not the same people who drive the vehicles – which are effectively a stealth tax on their users. And hotel operators who see their margins cut by the new competition from Airbnb are often global companies who repatriate their profits rather than spending them in the communities where their properties are located. Clearly there is a balance to be struck between necessary regulation on the one hand, and measures to stifle competition on behalf of vested interests on the other.

Digital disruption

An industry of authors and consultants has sprung up, offering to help businesses cope with this disruption. One of its leading figures is Peter Diamandis, who is also a co-founder of Silicon Valley's Singularity University. Diamandis talks about the Six Ds of digital disruption, arguing that the insurgent companies are:

Digitized, exploiting the ability to share information at the speed of light

Deceptive, because their growth, being exponential, is hidden for some time and then seems to accelerate almost out of control

Disruptive, because they steal huge chunks of market share from incumbents

Dematerialized, in that much of their value lies in the information they provide rather than anything physical, which means their distribution costs can be minimal or zero

Demonetized, in that they can provide for nothing things which customers previously had to pay for dearly

Democratized, in that they make products and services which were previously the preserve of the rich (like cellphones) available to the many.

The task for business leaders is to work out whether their industry can be disrupted by this sort of insurgent (hint: almost certainly yes) and whether they can do the disruption themselves rather than being left standing in rubble like Kodak.

Digital disruption isn't devastating only because it enables competitors to undercut your product and service price so dramatically. That cheapness also means there will be many more potential disrupters because the barriers to entry are disappearing. Small wonder that Monitor, the business consultancy established by Michael Porter to advise companies how to erect those barriers, went bankrupt.

Business leaders often know what they need to do: set up small internal teams (sometimes called skunk works[184]) of their most talented people to brainstorm potential disruptions and then go ahead and do the disrupting first. These teams need high-level support and freedom from the usual metrics of return on investment, at least for a while. The theory is fairly easy but putting it into practice is hard: most will need external help, and many will fail.

Of course the disrupters can also be disrupted. A service called LaZooz [185], based on blockchain technology, may provide serious competition for Uber. As computing and AI improve exponentially, the excitement and discomfort of disruption is only going to increase.

CHAPTER 13

EXPONENTIALS

The power of exponential growth

It is impossible to understand the scale of change that we face in the coming years without comprehending the astonishing impact of exponential increase. So here's a thought experiment. Imagine that you stand up and take 30 paces forward. You would travel around 30 metres. Now imagine that you take 30 exponential steps, doubling the distance travelled each step. Your first step is one metre, your second is two metres, your third is four metres, your fourth pace is eight metres, and so on.

How far do you think you would travel in 30 paces? The answer is, to the moon. In fact, to be precise, the 29th step would take you to the moon; the 30th step would bring you all the way back. That example illustrates not just the power of exponential increase, but also the fact that it is deceptive, and back-loaded.

Here is another illustration. Imagine that you are in a football stadium (either soccer or American football will do) which has been sealed to make it water-proof. The referee places a single drop of water in the middle of the pitch. One minute later she places two drops there. Another minute later, four drops, and so on. How long do you think it would take to fill the stadium with water? The answer is 49 minutes. But what is really surprising – and disturbing – is that after 45 minutes, the stadium is just 7% full. The people in the back seats are looking down and pointing out to each other that something curious is happening. Four minutes later they have drowned.[186]

The fact that exponential growth is back-loaded helps explain another phenomenon, known as Amara's Law, after the scientist Roy Amara. This says that we tend to over-estimate the effect of a technology in the short run and under-estimate the effect in the long run.[187] Similar observations have been made by Arthur C Clarke and Bill Gates.[188] How long are the short run and the long run? Bill Gates suggested two and ten years respectively. Science writer Matt Ridley suggests the long run is fifteen years.[189]

In the late 1990s, investors and others made radical decisions based on the belief that the internet was poised to change everything, and these decisions looked foolish in the early 2000s after the dotcom bubble burst. From the vantage point of 2020, however, they were simply premature, although of course in many areas of life, including investment, timing is crucial. Many earlier technologies, like steam power and electricity, have taken even longer to fulfil their promise. If 15-20 years is the right period, then we can expect extraordinary developments in 2022-2027, when the Big Bang in AI is 15-20 years old.

Often when someone says that x or y will never happen, they are only thinking about what may happen in the next five or ten years. The next time someone assures you that machines will never take human jobs, or that we will never have machines with general intelligence, ask them if they mean those things will not happen in a thousand years. Or a hundred years. And then ask them if they have taken the astounding impact of exponential growth into account. A lot of disagreements about what may happen in the future actually boils down to people not thinking about the time-frame. So with a lamentable lack of humility, I'm proposing a new rule, called Calum's Rule: "All forecasts about the impact of technology should specify the time frame."

No knees

People often talk about the "knee" of an exponential curve, the point at which past progress seems sluggish, and projected future growth looks dramatic. This is a misapprehension: there is no knee on an exponential curve. When you compare exponential curves plotted for ten and 100 periods of the same growth, they look pretty much the same. In other words, wherever you are on the curve, you are always at the beginning of the real growth phase: the past always looks horizontal and the future always looks vertiginous. We are only at the beginning of our AI journey, and we will always be there as long as exponential growth continues.

Moore's Law

In 1965, Gordon Moore was working for Fairchild Semiconductors. He published a paper observing that the number of transistors being placed on a chip was doubling every year. He forecast that this would continue for a decade, which his contemporaries considered extremely adventurous. Moore went on to co-found Intel in 1968, and a few years later (no-one remembers exactly when), a Caltech professor named Carver Mead coined the term Moore's Law.[190]

Moore's Law is not a law, but an observation which became a self-fulfilling prophecy – a target and a planning guide for the semiconductor industry, and for Intel in particular. It is generally taken to mean that the processing power of $1,000 of computer doubles every 18 months.

Exponential curves do not generally last for long: they are just too powerful. In most contexts, fast-growing phenomena start off slowly, pick up speed to an exponential rate, and then after a few periods they tail off to form an S-shaped curve. However exponentials can continue for many steps, and in fact each of us is one of them. Your body is composed of around 27 trillion cells, which were created by fission, or division – an exponential process. It required 46 steps of fission to create all of your cells. Moore's Law, by comparison, has had 37 steps in the 55 years of its existence.

No more Moore?

The death of Moore's Law has been declared and foretold for decades. In January 2017, Intel CEO Brian Krzanich said "I've heard the death of Moore's law more times than anything else in my career. And I'm here today to really show you and tell you that Moore's Law is alive and well and flourishing."[191] Peter Lee, a vice-president at Microsoft Research, joked that "there's a law about Moore's law: the number of people predicting the death of Moore's law doubles every two years."[192]

The truth is that Moore's Law is not dying, but changing – which is what it has always done. When Moore first described the phenomenon in 1965 he thought the doubling period was yearly. In 1975 he adjusted the period to two years, and shortly afterwards an Intel executive named David House noticed that the performance of individual transistors was also improving, so the period became 18 months. Until 2004, regular increases in the clock speeds of computer chips contributed a large part of their performance improvements. (See here[193] for an explanation of clock speeds, if you like that kind of thing.) Over-heating put a stop to this, and instead, chip manufacturers maintained Moore's Law by incorporating more processors, or "cores". Modern smartphones may have as many as eight, which means the processes they work on have to be broken down into pieces which are operated on in parallel.

Since 2007, Intel had pursued the development of

its chips with a "tick-tock" cadence. The tick represent-ed improvements in the manufacturing process, which enabled chip size to be reduced from 45nm to 32nm to 22nm to 14nm. The tock was improvements in the architecture. But in July 2015, Intel CEO Krzanich said that it was taking longer for the firm to cut the size of its transistors: "our cadence today is closer to 2.5 years than to 2." The new cadence was described by one ob-server as a move from tick-tock to tic-tac-toe, repre-senting process (tic), architecture (tac), and optimisa-tion and efficiency (toe).[194]

Apple's iPhone 11 uses 7nm chips, and the world's largest independent chip manufacturer (Taiwan's TSMC) announced the arrival of 5nm chips in April 2019.[195] For context, a human hair is 100,000 nanome-ters thick, and silicon atoms are around 0.2 nanometers across, so a 5nm structure is about 20 atoms wide.

New architectures

However successfully the chip manufacturers pro-long it, the existing architecture will reach its end point eventually, terminated by over-heating or quantum tunnelling. Researchers are hard at work on a number of technologies that will keep Moore's Law going.

Intel based its dominance on Central Processing Units, or CPUs. They are general purpose processors which can carry out many kinds of computation, but are not necessarily optimised for any of them. The re-quirements of video games prompted the development and refinement of Graphics Processing Units, or GPUs. These are very good at taking huge quantities of data and carrying out the same operation over and over again. It turns out that machine learning also benefits from their particular capabilities. CPUs and GPUs are often deployed in tandem.

The champion of GPUs is Nvidia, whose value rose ten-fold between 2014 and 2017, to more than $100 billion, as its chips proved invaluable for manufacturers of mobile phones as well as games consoles. Its rise was boosted still further by demand from bitcoin miners, but the share price halved in 2019, when the bitcoin bubble burst. In 2015 and 2016, Intel spent $17bn on acquiring Altera and Nervana, a hardware and a software company respectively, in order to catch up with Nvidia.

Still more specialised than GPUs are Google's TPUs, or Tensor Processing Units. These operate at very high speeds with lower precision than GPUs and CPUs, and are specifically designed for machine learning applications.[196]

Another new architecture being developed is 3D chips. Placing chips side-by-side delays the signals between them and causes bottlenecks as too many signals try to use the same pathways. These problems can be eased if you place the chips on top of each other, but this is hard. Silicon chips are fabricated at 1,800 degrees Fahrenheit, so if you manufacture one chip on top of another you will fry the one below. If you fabricate them separately and then place one on top of the other, you have to connect them with thousands of tiny wires. In December 2015, researchers from Stanford announced a new method of stacking chips which they called Nano-Engineered Computing Systems Technology, or N3XT. They claimed this was a thousand times more efficient than conventional chip configurations.[197]

Another approach is to combine memory chips with the traditionally separate processing chips, to reduce the amount of traffic between those two. In March 2016, scientists from IBM's TJ Watson Research Centre announced their belief that "resistive processing units", which combine CPU and memory on the same chip, could accelerate the processing of machine learning algorithms as much as 30,000 times.[198]

Yet another approach is to design chips specifically to implement neural networks. In the late 1980s, Carver Mead proposed the idea of computers based on the architecture of the brain, called neuromorphic computing. As with many developments in computing, it could not be realised at the time, but is now becoming a reality. In 2014 IBM produced a chip called TrueNorth, which comprised around a million silicon "neurons", each with 256 "synapses". As well as being very powerful, IBM claimed the chip is extremely energy-efficient.

Intel, still the world's largest chip manufacturer, is also working on neuromorphic computing, and in September 2017 it unveiled Loihi, which it described as the world's first self-learning neuromorphic chip: it mimics how the brain functions by learning from feedback received from its environment. As well as being powerful and energy-efficient, this type of chip is expected to be an effective platform for deep learning systems, and to require much less data than alternative deep learning platforms.[199]

Other companies developing neuromorphic chips include Qualcomm, Manchester University's SpiNNaker project, and BrainChip, which launched its Akida chip in October 2019.[200]

Towards quantum computing

Quantum computing, like nuclear fusion, is a technology which seems forever to be twenty years away from any practical application. In fact it has already arrived, but it is fantastically hard to implement – or even to understand. It is also fantastically expensive, and so far at least, its superiority over classical computers is limited to a small number of use cases. It is expected to create vulnerabilities in some of the encryption techniques currently used by banks and others, so researchers are hard at work on quantum-resistant cryptography.[201]

Quantum computing is based on the idea that while classical computers use bits (binary digits) which are either on or off, quantum bits (qubits) can be both on and off at the same time – known as superposition. This enables them to carry out a number of different calculations at once. It also employs the phenomenon of quantum entanglement, or "spooky action at a distance", as Einstein derisively described it.

Google, IBM and Microsoft are all investing large sums in pursuit of "quantum supremacy", a term introduced by professor John Prentiss in 2012 to mean the ability to perform calculations which classical computers would not be able to. In October 2019 Google declared it had achieved quantum supremacy by generating a result which would have taken the world's fastest supercomputers 10,000 years, although IBM claimed that this was not yet true quantum supremacy.[202]

In April 2020, a researcher reported that "The Google plan is roughly to build a million-qubit system in about ten years".[203]

Moore's Law for software

Moore's Law is generally considered to be a hardware phenomenon. Chris Bishop, the head of Microsoft Research in the UK, argues that we are seeing the start of something else: "I think we're seeing ... a similar, singular moment in the history of software... The rate limiting step now is ... the data, and what's really interesting is the amount of data in the world is – guess what – it's growing exponentially! And that's set to continue for a long, long time to come."[204]

There's a Cambrian explosion going on, and some people say that Moore's Law understates how fast things are moving.

So what?

Assuming Moore's Law continues, the machines we have in 10 years time will be more than 120 times more powerful than the ones we have today. In 20 years they will be 8,000 times more powerful. And in 30 years time – in 2050 – they will be a million times more powerful. It is hard to exaggerate the significance of this improvement in performance. This is not business as usual.

CHAPTER 14

IT'S NOT THE FOURTH INDUSTRIAL REVOLUTION

Klaus Schwab, founder and executive chairman of the World Economic Forum which hosts the annual meeting of the global elite in Davos, describes the arrival of ubiquitous, mobile supercomputing, intelligent robots and self-driving cars as the fourth industrial revolution.[205] In 2017 he published a book of that name which contains much to be admired, but the nomenclature is confusing and unhelpful.

This is not the first time someone has claimed that a new technology heralded the fourth industrial revolution – in fact it is at least the sixth. (The other five, since you ask, were atomic energy in 1948, ubiquitous electronics in 1955, computers in 1970, the information age in 1984, and finally, nanotechnology.)[206]

The important problem with the label is that it greatly under-states the importance of what is happening. The transition to an AI-centric world is not the fourth industrial revolution.

Instead, it's the fourth human revolution[207]

The first great revolution to transform the nature of being human was the cognitive revolution, which took place around 50-80,000 years ago, and in which we acquired the communication skills which turned a low-level scavenger into the most fearsome predator on the planet. (Anthropologists argue about when it happened and how long it took.)

Next came the agricultural revolution which turned foragers into farmers. It happened in different parts of the world at different times from around 12,000 years ago. It gave us mastery over animals, and generated food surpluses which allowed our population to grow enormously. It enabled the rise of cities, which have been described as engines of innovation. It made the lives of most individual humans considerably less pleasant, but it greatly advanced the species.

The third great revolution was the industrial one, which in many ways gave us mastery of the planet. The product of capitalism, the enlightenment, and the discovery of the scientific method, it ended the perpetual tyranny of famine and starvation, and brought the majority of the species out of the abject poverty which had been the fate of almost every human before. For most people in the developed world it created lifestyles which would have been the envy of kings and queens in previous generations.

The information revolution is our fourth great transformation, and it will have even more profound impacts than any of its predecessors. AI is the most powerful part of this information revolution. Most people have yet to grasp the significance of the changes heading our way. We need to change this if we are to navigate the transition successfully. Under-stating the importance of the information revolution by demoting it to a sub-set of the industrial one is unhelpful.

CHAPTER 15

SINGULARITIES

Origins

In maths and physics, the term "singularity" means a point at which a variable becomes infinite. The classic example is the centre of a black hole, where the gravitational field becomes infinite, and the laws of physics cease to operate. When you reach a singularity, the normal rules break down, and the future becomes even harder to predict than usual.

The term was first applied to human affairs in the 1950s by the polymath John von Neumann, one of the founding figures of modern computing. In a eulogy published in 1958, the Polish mathematician (and inventor of the Monte Carlo method of computation) Stanislaw Ulam wrote of a conversation he had with von Neumann which "centred on the ever-accelerating progress of technology and changes in the mode of human life, which gives the appearance of approaching some essential singularity in the history of the race beyond which human affairs, as we know them, could not continue."[208]

The concept was picked up by scientist and science fiction author Vernor Vinge, who argued in a 1993 paper that sometime between 2005 and 2030, "superhumanly intelligent" and conscious computers would be created, and that the enormous changes to human life that this would produce could be termed a "technological singularity".[209]

Ray Kurzweil

The best-known proponent of the idea that humanity is approaching a singularity is Ray Kurzweil. A successful inventor and businessman with a string of ventures in speech recognition software, optical character recognition systems, and music synthesisers, Kurzweil has also written a series of highly influential books. In The Age of Intelligent Machines (1990), the Age of Spiritual Machines (1999), and The Singularity is Near (2006), he argues that Moore's Law is a special case of a universal principle he calls the Law of Accelerating Returns. He believes that this law means that humanity will create an artificial general intelligence in 2029, and this will lead to a singularity in 2045, after which humans will merge with machines and become immortal and godlike.

Not surprisingly, this is a lot for many people to swallow, and he has fierce critics. People as diverse as science fiction author Neal Stephenson and robotics researcher Rodney Brooks dismiss Kurzweil's ideas as risible, and many people have labelled them as pseudo-religious. The term "singularity" became associated with a naïve belief that technology, and specifically a superintelligent AI, would magically solve all our problems, and that everyone would live happily ever after. Because of these quasi-religious overtones, the singularity has frequently been satirised as "rapture for nerds", by analogy with the fundamentalist Christian idea that believers will be taken up into Heaven prior to the second coming of Christ. Kurzweil's eccentricities, like his daily diet of hundreds of pills, and his somewhat robotic speech pattern, probably don't help.

Since Kurzweil has been in the predictions business for several decades, various people have tried to assess his accuracy. Some of his forecasts were both bold and accurate, including the victory of a chess computer, the collapse of the Soviet Union, and the growth of the world wide web. Kurzweil himself claimed in October 2010 that of 147 predictions, 115 were entirely correct, 29 were essentially or partially correct and only three were actually wrong.[210] Others have been far less charitable: John Rennie argued that the predictions are often vague enough as to be un-falsifiable.[211] Others have argued that his main flaw is optimistic timing, and that many of his forecasts happen about a decade late.

Kurzweil has fans as well as critics, including the founders of Google, who hired him as a director of engineering there in 2012. (I visited the Googleplex the following year, and I asked my guide which of the many buildings he worked in. She looked him up and replied "in building 42". She appeared not to know the significance of that, but I imagine Kurzweil's employers did.[212]) One of the projects Kurzweil has been working on at Google is Smart Reply, a feature in Gmail, Google's browser-based email service, that suggests responses to emails in your inbox.

Sticking with singularity

Whatever the merits or otherwise of Kurzweil's specific predictions, there is no doubt that the term "singularity" has become tarnished by accusations of pseudo-religious enthusiasm. Thus, even the Singularity University, a private educational and training institution co-founded by Kurzweil and X-Prize founder Peter Diamandis to help people apply exponential principles to solve humanity's big problems, seems almost allergic to saying anything about the singularity.

This is a shame. Kurzweil's books have probably done more than anything else to alerting people to the scale of the changes heading our way, including the possibility that AGI and superintelligence may be decades away rather than millennia away. And as a superlative for change, the term "singularity" is a graphic one with a perfectly respectable intellectual origin. Absent such a term, we have to resort to usages like "complete transformation" and "total change", which are dreary, and easy to debase by application to more pedestrian developments. We could talk about humanity going through phase transitions, like ice melting to become water, or water boiling to become steam, but that seems too deterministic, with outcomes dictated by the laws of physics and chemistry.

Multiple singularities

It has been suggested that there cannot be two singularities because they would no longer be singular. This is nonsense. There are believed to be black holes at the centre of every galaxy, and there are believed to be a hundred billion or so galaxies in the observable universe. Each black hole contains a singularity. Clearly a singularity need not be unique.

Distinct from the question whether there could be more than one singularity, there are rival definitions. Eliezer Yudkowski is an American AI researcher, and co-founder of the Machine Intelligence Research Institute, and he has been thinking about these things longer than most. He identifies three main categories of usage: accelerating change, event horizon, and intelligence explosion.[213] If he ever reads this book, I fear it may irritate Yudkowski, as I'm not using the word in strict accordance with any of these categories, but simply as a superlative for change driven by technology. But I think I'm using it in the sense that von Neumann originally intended.

Singularity riffs

When an idea becomes popular, and perhaps goes viral, it attracts puns and derivatives. Disney's 2008 film "Wall-E" depicted a possible future in which robots do all the work and humans become decadent and lazy. Wags have labelled this the "sofalarity".

In 2010, a journalist suggested that nearly-perfect speech recognition technology would be available soon. He dubbed this the "speakularity".[214]

Two singularities

The next century or two are likely to witness two singularities. The technological singularity is the arrival of artificial general intelligence (AGI), which leads to superintelligence. If and when this happens, it will be the most important thing ever to happen to humanity. It will also be either the best or the worst thing ever to happen to us. I cover this at length in my book "Surviving AI".

The economic singularity is likely to come sooner. It is when we have to change the basis of our economies because we have to admit that technological unemployment is real, and that many or most humans will not be able to earn a living from work. That is the subject of the rest of this book.

PART THREE

THE ECONOMIC SINGULARITY

CHAPTER 16

THE HISTORY OF AUTOMATION

The industrial revolution

For a process that began hundreds of years ago, the start date for the industrial revolution is surprisingly controversial. Historians and economists cannot even agree how many industrial revolutions there have been: some say there has been one revolution with several phases, others say there have been two, and others say more. It has become fashionable to say that we are entering a fourth industrial revolution, but I argued in chapter 12 that this is very unhelpful.

The essence of the industrial revolution was the shift from manufacturing goods by hand to manufacturing them by machine, and the harnessing of better power sources than animal muscle. So a good date for its beginning is 1712, when Thomas Newcomen created the first practical steam engine for pumping water. For the first time in history, humans could generate more power than muscles could provide - wherever they needed it.

The replacement of human labour by machines in manufacturing dates back considerably earlier, but they were powered by muscles or by wind or water. In the 15th century, Dutch workers attacked textile looms by throwing wooden shoes into them. The shoes were called sabots, and this may be the etymology of the word "saboteur". A century later, around 1590, Queen Elizabeth (the First) of England refused a patent to William Lee for a mechanical knitting machine because it would deprive her subjects of employment.

In the second half of the 18th century, the Scottish inventor James Watt teamed up with the English entrepreneur Matthew Boulton to improve Newcomen's steam engine so that it could power factories, and make manufacturing possible on an industrial scale. At the same time, iron production was being transformed by the replacement of charcoal by coal, and "canal mania" took hold, as heavy loads could be transported more cheaply by canal than by road or sea.

Later, in the mid-19th century, steam engines were improved sufficiently to make them mobile, which ushered in the UK's "railway mania" of the 1840s. Projects authorised in the middle years of that decade led to the construction of 6,000 miles of railway – more than half the length of the country's current rail network. Other European countries and the USA emulated the UK's example, usually lagging it by a decade or two.

Toward the end of the 19th century, Sir Henry Bessemer's method for converting iron into steel enabled steel to replace iron in a wide range of applications. Previously, steel had been an expensive commodity, reserved for specialist uses. The availability of affordable steel enabled the creation of heavy industries, building vehicles for road, rail, sea and later the air.

As the 20th century arrived, oil and electricity provided versatile new forms of power and the industrial world we recognise today was born. The changes brought about by these technologies are still in progress.

In summary, we can identify four phases of the industrial revolution:

1712 onwards: the age of primitive steam engines, textile manufacturing machines, and the canals

1830 onwards: the age of mobile steam engines and the railways

1875 onwards: the age of steel and heavy engineering, and the birth of the chemicals industry

1910 onwards: the age of oil, electricity, mass production, cars, planes and mass travel.

From an early 21st-century standpoint, it seems entirely natural that the industrial revolution took off where and when it did. In fact it is something of a mystery. Western Europe was not the richest or most advanced region of the world: there were more powerful empires in China, India and elsewhere. There is still room for debate about whether the technological innovations came about in England at that time because of the cultural environment, the legal framework, or the country's fortuitous natural resources. Fascinating as these questions are, they need not detain us.

The information revolution

Even though the industrial revolution is still an on-going process, there is general agreement that we are now in the process of an information revolution. There is less consensus over when it began or how long it is likely to continue.

The distinguishing feature of the information revolution is that information and knowledge became increasingly important factors of production, alongside capital, labour, and raw materials. Information acquired economic value in its own right. Services became the mainstay of the overall economy, pushing manufacturing into second place, and agriculture into third.

One of the first people to think and write about the information revolution and the information society was Fritz Machlup, an Austrian economist. In his 1962 book, The production and distribution of knowledge in the United States, he introduced the notion of the knowledge industry, by which he meant education, research and development, mass media, information technologies, and information services. He calculated that in 1959, it accounted for almost a third of US GDP, which he felt qualified the US as an information society.

Alvin Toffler, author of the visionary books Future Shock (1970) and The Third Wave (1980), argued that the post-industrial society has arrived when the majority of workers are doing brain work rather than personally manipulating physical resources. Working in the service sector was taken as a rough-and-ready proxy for this, and services first accounted for 50% of US GDP shortly before 1940,[215] and they first employed the majority of working Americans around 1950.

Overlapping revolutions

We have seen that the start and end dates of the economic revolutions (agricultural, industrial and information) are unclear. What's more, they can overlap, and sometimes re-ignite each other.

An example of this overlap is provided by the buccaneers who preyed on Spanish merchant shipping en route to and from Spain's colonies in South America during the 17th century. (Some of these buccaneers were effectively licensed in their activities by the English, French and Dutch crowns, which issued them with "letters of marque". This ceased when Spain's power declined toward the end of the century, and the buccaneers became more of a nuisance than a blessing to their former sponsors.) When a buccaneer raiding party boarded a Spanish ship the first thing they would look for and demand was the maps. Charts – a form of information which improve navigation – were actually more valuable than silver and gold.[216]

An example of one revolution re-igniting another is that the industrial revolution enabled the mechanisation of agriculture, causing a second agricultural revolution, making the profession of farming more effective and more efficient. The information revolution does the same, providing farmers with crops that are more resilient in the face of weather, pests and weeds, and allowing them to sow, cultivate and harvest their crops far more accurately with satellite navigation.

The automation story so far

The particular aspect of the industrial and information revolutions which concerns us in this part of this book is automation. Perhaps the clearest example of automation destroying jobs is the mechanisation of agriculture, a sector which accounted for 50% of US employment in 1870,[217] and only 1.5% by 2004.[218]

Many of the people who quit farm work moved to towns and cities to take up other jobs because they were easier, safer, or better paid. Others were forced to find alternative employment because they could not compete with the machines. This process caused much suffering to individuals, but overall, the level of employment did not fall in the long run, and society became far richer – both in total and on average. More than one new job was created for every job that was lost. In his book, "A World Without Work", Daniel Susskind describes this as the "compensatory effect" overcoming the "substitutive effect".

As machines replaced muscle power on the farm, humans had other skills and abilities to offer. Factories and warehouses took advantage of our manual dexterity and our ability to carry out a very broad range of activities. Office jobs used our cognitive ability. We turned our hands (often literally) to more value-adding work: you could say that we climbed higher up the value chain.

One-trick ponies

While the mechanisation of agriculture was a good news story for humans, it didn't work out so well for the horse, which had nothing to offer beyond muscle power. 1915 was when the US reached "peak horse", with a population of 21.5m – most of them pulling vehicles on roads and farms.[219] By the early 1950s, the motor car and the tractor had replaced those horse jobs and the horse population collapsed to just 2 million. Sadly, many of the disappearing horses did not get to live out their natural lives on grassy pensions, but were sent off to slaughterhouses: their final job was being dog food. It would be hard to devise a more graphic illustration of automation causing lasting widespread unemployment.[220]

Artificial intelligence systems and their peripherals, the robots, are increasingly bringing flexibility, manual dexterity, and cognitive ability to the automation process. One of the big questions addressed in this part of this book is: as computers take over the role of ingesting, processing and transmitting information, will there be anywhere higher up the value chain for humans to retreat to? In other words, can we avoid playing the role of the horse in the next wave of automation? Are we approaching "peak human" in the workplace?

Mechanisation and automation

What went on in farms was mechanisation rather than automation, and the distinction is important. Mechanisation is the replacement of human and animal muscle power by machine power; a human may well continue to control the whole operation. Automation means that machines are controlling and overseeing the process as well: they continuously compare the operation to pre-set parameters, and adjust the process if necessary.

Although the word "automation" was not coined until the 1940s by General Electric,[221] this description applies pretty well to the operation of 19th-century steam engines once James Watt had perfected his invention of governors – devices which control the speed of a machine by regulating the amount of fuel they are supplied. Automated controllers which were able to modify the operation more flexibly became increasingly common in the early 20th century, but the start-stop decisions were still normally made by humans.

In 1968 the first programmable logic controllers (PLCs) were introduced.[222] These are rudimentary digital computers which allow far more flexibility in the way an electrochemical process operates, and eventually general-purpose computers were applied to the job.

The advantages of process automation are clear: it can make an operation faster, cheaper, and more consistent, and it can raise quality. The disadvantages are the initial investment, which can be substantial, and the fact that close supervision is often necessary. Paradoxically, the more efficient an automated system becomes, the more crucial the contribution of the human operators. If an automated system falls into error it can waste an enormous amount of resources and perhaps cause significant damage before it is shut down.

Let's take a look at how automation has affected some of the largest sectors of the economy.

Retail and "prosumers"

Retail is a complicated business and there have been attempts to automate many of the processes required to get goods from supplier to customer, and payment from customer to supplier. Demand forecasting, product mix planning, purchasing, storage, goods handling, distribution, shelf stacking, customer service and many other aspects of the business have been automated to varying extents in different places and at different times.

The retail industry has also given us the clearest examples of another, associated phenomenon known as "prosumption", a term coined in 1980 by Alvin Toffler. At the same time as organisations automate many of their processes, they enlist the help of their customers to streamline their operations. In fact, they get their customers to do some of the work that was previously done for them. The reason why consumers accept this (indeed welcome it) is that the process speeds up, and becomes more flexible – more tailored to their wishes.

Toffler first described this process in FutureShock, and in The Third Wave he defined a "prosumer" as a consumer who is also involved in the production process. Where once people were passive recipients of a limited range of goods and services designed or selected by retailers, he foresaw that we would become increasingly involved in their selection and configuration.

Perhaps the simplest example of what he meant is the purchase of gasoline. This dangerous substance was traditionally dispensed by pump attendants, but Richard Corson's invention of the automatic shut-off valve enabled the job to be taken over by customers. Nowadays most consumers in developed countries dispense their own gasoline at self-service pumps. This saves money for the retailer and time for the consumer.[223]

Supermarkets have often led the way in automation and prosumption because they are owned by massive organisations with the budgets and the sophistication to invest in the systems needed. Decades ago, what marketers call fast-moving consumer goods (or fmcg - foods, toiletries, etc.) were requested one at a time by the shopper at a counter and fetched individually by the shopkeeper or his assistant. As these general merchandise firms grew bigger and more sophisticated they built large stores where shoppers fetched their own items, and presented them for processing at checkouts, like components on a car assembly line. Later on, self-service tills were installed, where shoppers could scan the bar codes of their goods themselves, speeding up the process considerably.

At each stage of this evolution, the involvement of the consumer in selecting and transporting each item increases, and the requirement for shop staff involvement reduces. This latter effect is disguised because, as society gets richer, people buy many more items, so the store needs more staff even though their involvement in each individual item is less.

Online shopping is perhaps the ultimate prosumer experience. Consumer reviews replace the retailer's sales force, and its algorithms do the up-selling.

Food service

The automation of service in fast food outlets seems to have been just around the corner for decades. Indeed, elements of it have been a reality for years in Oriental-style outlets like Yo, Sushi!, but it has so far failed to spread to the rest of the sector. There are several reasons for this, including the relatively low labour cost of people working in fast food outlets, and the need for every single purchase to be problem-free, and if not, for there to be a trained human on hand to solve any problem immediately. If a hands-free wash basin fails 5% of the time it is no big deal, but it would be a very big problem if 5% of meals were inedible, or delivered to the wrong customer.

A combination of factors is poised to overcome these hurdles. Increases to the cost of labour caused by rising minimum wage legislation, declining costs of the automated technology, greater cultural acceptance of interacting with machines, and above all, the improved performance of the automated technology. It is increasingly flexible, and it goes wrong less often.

Manufacturing

Car manufacturing has traditionally incurred relatively high labour costs. The work involves a certain amount of physical danger, with heavy components being transported, and metals being cut and welded. It is also a sector where a lot of the operations can be precisely specified and were highly repetitive. These characteristics make it ripe for automation, and the fact that the output (cars) are high-value items means that investment in expensive automation systems can be justified. Around half of all the industrial robots in service today are engaged in car manufacturing.[224]

Despite the lingering effect of the great recession, sales of industrial robots grew at 16% a year from 2010 to 2016, when 254,000 units were sold worldwide. In 2018, the latest year for which data is available from the IFR (International Federation of Robotics), 380,000 industrial robots were shipped. China became the biggest market in 2013, and 75% of all robot sales were made in these five countries: China, the Republic of Korea, Japan, the United States, and Germany.[225]

Until recently, the industrial robots used in car manufacturing (and elsewhere) were expensive, inflexible, and dangerous to be around. But the industrial robotics industry is changing: as well as growing quickly, its output is getting cheaper, safer and far more versatile.

One example was Baxter, a 3-foot tall robot (6 feet with his pedestal) from Rethink Robotics. The brainchild of Rodney Brooks, an Australian roboticist who used to be the director of the MIT Computer Science and Artificial Intelligence Laboratory (CSAIL), Baxter was much less dangerous to be around. By early 2015, Rethink had received over $100m in funding from venture capitalists, including the investment vehicle of Amazon founder Jeff Bezos. Baxter was intended to disrupt the industrial robots market by being cheaper, safer, and easier to programme. He was certainly cheaper, with a starting price of $22,000. He was safer because his arm and body movements were mediated by springs, and he carried an array of sensors to detect the presence nearby of squishy, fragile things like humans. He was easier to programme because an operator could teach him new movements simply by physically moving his arms in the intended fashion.

Sales did not pick up as expected, and in December 2013, Rethink laid off around a quarter of its staff. In March 2015 it introduced a smaller, faster, more flexible robot arm called Sawyer. It can operate in more environments than Baxter, and can carry out more intricate movements. It is slightly more expensive, at $29,000. This was not enough to save the company, and in 2018, Rethink ceased trading, and most of its assets were acquired by a German distributor, the Hahn Group.[226]

Developing robots is hard, and Rethink will not be the last potential star to fall to earth. But progress is being made, and one of the places where this is evident is warehouses.

Warehouses

As we saw in chapter 11, another of Jeff Bezos' robotic investments has been far more successful. Kiva Systems was established in 2003, and acquired by Amazon in 2012. It produces robots which collect goods on pallets from designated warehouse shelves and deliver them to human packers in the bay area of the warehouse. In June 2019, Amazon had 200,000 Kiva robots in its warehouses.[227]

Designing and manufacturing robots is challenging, and the economics even more so. But the direction of travel is clear, and enormous investments are being made.

Secretaries

Most of the jobs which have been automated so far (farm labourers, lift attendants, petrol pump attendants, and so on) involved manual work. But some occupations which depend almost entirely on cognitive skills have also been largely automated out of existence: secretaries are an example. In the 1970s, most managers had secretaries, and generally did little work on computers themselves. In 1978, "secretary" was the most common job title in 21 of the 50 US states. Today, many managers spend much of their day staring at computer screens, and "secretary" is the most common job in only 4 US states.[228]

Early days

There are has been a lot of debate about whether automation has caused unemployment so far - for humans. (As discussed above, it certainly has for horses.) A study published in March 2017 by Acemoglu and Restrepo claimed that it has, but on a modest scale.[229] Adjusting for the effects of globalization, one extra robot per thousand workers decreased employment by 5.6 workers and cut wages by around 0.5 percent. The authors noted that there are still relatively few robots in the economy. And we are only at the beginning.[230]

Ned Ludd

A person can have a big impact on society without going to the trouble of actually existing. In 1779, Ned Ludd was a weaver in Leicester who responded to being told off by his father (or perhaps his employer) by smashing a machine. Or maybe he wasn't – the truth is, we don't know. He certainly wasn't the leader of an organised group of political protesters. Nevertheless, in the decades following his alleged outburst, his name was commonly used to take the blame for an accident or an act of vandalism.

As Britain pioneered the industrial revolution in the late 18th and early 19th centuries, many of its people attributed their economic misfortune to the introduction of labour-saving machines. They were no doubt partly correct, although poor harvests, and the Napoleonic Wars against France were also to blame. There was a short-lived phenomenon of organised protest under the banner of Luddism in Nottingham in 1811-13: death threats signed by King Ludd were sent to machine owners.

The government responded harshly, with a show trial of 60 men (many of them entirely innocent) in York in 1813. Machine breaking was made a capital offence. Riots continued sporadically, notably in 1830-31, when the Swing rioters in southern England attacked threshing machines and other property. Around 650 of them were jailed, 500 were sent to the penal colony of Australia, and 20 were hanged.[231]

The Luddite fallacy

The Luddites, and other rioters, were not making a general economic or political observation that the introduction of labour-saving machinery inevitably causes mass unemployment and privation. They were simply protesting against their own dire straits, and demanding urgent help from the people who were obviously benefiting.

It is therefore slightly unfair to them that the term "Luddite fallacy" has become a pejorative term for the mistaken belief that technological development necessarily causes damaging unemployment. (Although, given the hunger they were experiencing, they would probably regard the slur as the least of their problems.)

The Luddite fallacy pre-dates the industrial revolution, and has taken in quite a few heavyweight thinkers down the years. As long ago as 350BC, the Greek philosopher Aristotle observed that if automata (like the ones said to be made by the god Hephaestus) became so sophisticated that they could do any work that humans do, then workers - including slaves - would become redundant.[232]

During the early 19th century, when the industrial revolution was in full swing, most members of the newly-established social science of economics argued that any unemployment caused by the introduction of machinery would be resolved by the growth in overall economic demand. But there were prominent figures who took the more pessimistic view, that innovation could cause long-term unemployment. They included

Thomas Malthus, John Stuart Mill, and the most respected economist of the time, David Ricardo.[233]

The Luddite fallacy and economic theory

There are two main reasons why it has been reasonable to reject the Luddite fallacy up until now. The first reason is economic theory. Companies introduce machines because they increase production and cut costs. This increase in supply builds up the wealth in the economy as a whole, and hence the demand for labour.

Say's Law, named after French economist Jean-Baptiste Say, holds that supply creates its own demand, and Say argued that there could not be a "general glut" of any particular goods. Of course we do see gluts in sectors of the economy, but an adherent of Say's Law would argue these are the unintended consequences of interventions in free markets, usually by governments. This law became a major tenet of classical economics, but it was rejected emphatically by British economist John Maynard Keynes, and is not widely accepted today.

But many economists would accept a broader interpretation of the law which states that reducing the cost of a significant product or service will free up money which was previously allocated to it. This money can then be spent to buy more of the item, or other items, thereby raising demand generally, and creating jobs. This assumes that the money freed up is not spent on expensive assets that generate no employment, or invested in companies that employ very few people.

Economists also point out that the Luddite fallacy

depends on a misapprehension about economics called the "lump of labour fallacy", which is the idea that there is a certain, fixed amount of work available, and if machines do some of it then there is inevitably less for humans to do. In fact, economies are more organic and more flexible: they respond to shifts, and innovate to grow. New jobs are created as old ones disappear and the former outnumber the latter.

The Luddite fallacy and economic experience

The second reason to reject the Luddite fallacy hitherto is the claim that history has proved it to be wrong. A great deal of machinery has been deployed since the start of the industrial revolution, and yet there are more people working today than ever before. If the Luddite fallacy was correct, the argument goes, we would all be unemployed by now.

A study published in August 2015 by the business consultancy Deloitte analysed UK census data since 1871 and concluded that far more jobs have been created than destroyed by technology in that time.[234] Furthermore, the study argued that the quality of the jobs has improved. Where people used to do dangerous and gruelling jobs on the land, and hundreds of thousands used to do the work now done by washing machines, many more Britons are now employed in caring and service jobs. In the last two decades alone there has been a 900% rise in nursing assistants, a 580% increase in teaching assistants, and a 500% increase in bar staff – despite the closure of so many of the country's pubs. (The authors refrained from commenting on the finding that the number of accountants has doubled.)

So most economists would agree that in the long run, the Luddite fallacy is just that – a fallacy. But in the short run the Luddites had a point. Economists think that in the first half of the 19th century, wages failed to keep pace with increases in labour productivity. An economist named Arthur Bowley observed in the early 20th century that the share of GDP which goes to labour is generally roughly equal to that which goes to capital,[235] but in the first half of the 19th century, the share of national income taken by profit increased at the expense of both labour and land. The situation changed again in the middle of the century and wages resumed their normal growth in line with productivity. It may be that the slippage in wages was necessary and inevitable to enable enough capital to be accumulated to fuel the investment in technological change.

The period in the early 19th century when wage growth lagged productivity growth is known as the Engels pause, after the German political philosopher Friedrich Engels, who wrote about it in the 1848 "Communist Manifesto", which he co-authored with Karl Marx. The effect ceased at pretty much the same time as he drew attention to it, which may explain why it is not better known.[236]

The ATM myth

In an engaging TED talk recorded in September 2016[237], economist David Autor points out that in the 45 years since the introduction of Automated Teller Machines (ATMs), the number of human bank tellers doubled from a quarter of a million to half a million. He argues that this demonstrates that automation does not cause unemployment – rather, it increases employment.

He says ATMs achieved this counter-intuitive feat by making it cheaper for banks to open new branches. The number of tellers per branch dropped by a third, but the number of branches increased by 40%. The ATMs replaced a big part of the previous function of the tellers (handing out cash) but the tellers were liberated to do more value-adding tasks, like selling insurance and credit cards.

This story about ATMs has become something of a meme, popular with people who want to believe that technological unemployment is not going to be a thing.

Unfortunately it is almost certainly not true. The increase in bank tellers was not due to the productivity gains afforded by the ATMs. According to an analysis by finance author Erik Sherman, the increase was mostly due instead to a piece of financial deregulation, the Riegle-Neal Interstate Banking and Branching Efficiency Act of 1994, which removed many of the restrictions on opening bank branches across state lines. Most of the growth in the branch network occurred after this Act was passed in 1994, not before it.238

This explains why teller numbers did not rise in the same way in other countries during the period. In the UK, for instance, retail bank employment just about held steady at around 350,000 between 1997 and 2013[239], despite significant growth in the country's population, its wealth, and its demand for sophisticated financial services.

In any case, the example of ATMs tells us little about the likely future impact of cognitive automation: they are pretty dumb machines.

Is it different this time?

Mechanisation and automation has displaced workers on a huge scale since the beginning of the industrial revolution. It has imposed considerable suffering on individuals, but has led to greater wealth and higher levels of (human) employment overall. The question today is whether that will always be true. As machines graduate from offering just physical labour to offering cognitive skills as well, will they begin to steal jobs that we cannot replace? If the early 20th century saw "peak horse" in the workplace, will the first half of the 21st century see "peak human"?

In other words, is it different this time?

CHAPTER 17

WHAT PEOPLE DO

Jobs, not work

It is important to distinguish between jobs and work. Physicists define work as the expenditure of energy to move an object,[240] but what we mean by it here is the application of effort to achieve a task. That effort could be physical, mental, or both. Work can be instigated by an employer, but it can also be unpaid and purely personal: building or decorating a home, pursuing a hobby, or a volunteer project.

the purposes of this discussion, a job is paid work. It might be a salaried occupation with a single, stable employer, or it could involve self-employment, or freelance activity. Your job is the way you participate in the economy, and earn the money to buy the goods and services that you need to survive, and (hopefully) enjoy a good standard of living.

If a machine carries out a job cheaper, better and

faster than a human can, there is no point asking a human to do the job instead. The human will have to look for some other way to generate an income.

Jobs and tasks

In November 2015, the consultancy McKinsey contributed to the debate about technological unemployment with the observation that jobs are comprised of bundles of tasks, and that that machines often don't acquire the ability to automate entire jobs in one fell swoop. Instead they become able to automate certain of the tasks which people in those jobs perform.[241]

So what are these tasks? What exactly is it that people do for a living?

The three-sector economic model

The model of the economy which has been prevalent for the last century recognises three main sectors. The primary sector extracts raw materials, and includes agriculture and mining. The secondary sector builds and manufactures. The tertiary sector provides services. The model is sometimes extended to five sectors by adding the quaternary (fourth) sector, which includes education and research, and the quinary (fifth) sector, which includes high-level decision-making in government and industry.

Prior to the industrial revolution, most working people were employed in the primary sector, and this is still true in the least developed economies. Countries with median national incomes are dominated by the secondary sector, and the richest countries by the tertiary sector. In the UK, for instance, service industries account for 78% of GDP, with manufacturing accounting for 15%, construction for 6%, and agriculture less than 1%.[242]

Blue collar, white collar

The term "white collar" was first used in 1910 in a Nebraska newspaper, to mean someone engaged in cognitive work, probably sitting at a desk. Fourteen years later, an Iowa newspaper coined the term "blue collar" to mean someone doing manual work.[243]

A range of other collar colours have since been added:

pink collar, for service jobs like waiters and retail assistants, which used to be held primarily by women

gold collar, for knowledge workers like highly trained engineers who operate in manual labour environments

grey collar, for workers past the retirement age

orange collar, for prison inmates

no collar, for artists and "free spirits"

Traditionally, blue collar, or manual jobs, were associated with the primary and secondary sectors (extraction and manufacturing). White collar, or cognitive jobs, were associated with the tertiary sector (services). But in reality, both purely manual and purely cognitive work are found in all three sectors, along with work which involves both manual and cognitive abilities.

Manual tasks

When trying to automate manual tasks, we encounter Moravec's Paradox. This, as we noted in chapter 6, explains that getting machines to do things that we find hard (like playing chess at grandmaster standard) can be relatively easy, while getting them to do things that we find easy (like opening a door) can be hard.

Many jobs involving manual dexterity, for instance, or the ability to traverse un-mapped territory, are currently hard to automate. But as we will see in later chapters, that is changing.

Cognitive tasks

Many people spend their whole day at work dealing with information: obtaining it, processing it, and passing it on to others. This is as true for people in the manufacturing, construction and agricultural sectors as it is for people in the service sector.

Obtaining information can involve carrying out research, asking colleagues, looking online or occasionally in books, or coming up with an original idea – which itself usually involves combining two or more ideas from elsewhere.

Processing information can mean checking its accuracy or relevance, determining its importance relative to other pieces of information, making a decision about it or performing some kind of calculation on it. Passing information on is increasingly achieved electronically, for instance by email or online work flow systems.

Obtaining, processing and passing on information

can be solitary endeavours, or they can be carried out collaboratively with other people. Almost by definition, the solitary tasks could be carried out by a machine which possesses human-level (or above) ability to understand speech, recognise images, and a modicum of common sense.

Working with other people

Collaboration with other humans can also take many forms: brainstorming with colleagues; preparing for and negotiating a deal which will yield benefit to both sides but maximise your own; pitching an idea to a self-important, unimaginative and prickly boss; coaching a subordinate who has talent, but is also naïve. These tasks are far harder for a machine to emulate.

But for how long? Plenty of interactions with humans can be successfully automated today. Most people prefer withdrawing cash from ATMs than dealing with human cashiers, and many choose to use automated checkouts in retail stores. The centre of gravity of the entire retail industry is shifting online, where consumers deal with machines (via websites) rather than humans. Globally, the penetration of e-commerce was 14% in 2019, and it is rising fast.[244]

This does not mean that humans are becoming anti-social – far from it. Merely that we like to be able to choose for ourselves when we interact in a leisurely manner with another human, and when we have a brisk and efficient interaction (with a machine or a human) in order to conduct some business.

Machines are surprisingly good at some tasks which appear at first sight to require a human touch. In chapter 20 we will meet Ellie, a machine therapy system developed by DARPA, the research arm of the US military, which has proved surprisingly effective at diagnosing soldiers with post-traumatic stress disorder.

The gig economy and the precariat

As McKinsey noted, jobs can be analysed into tasks, some of which can be automated with current machine intelligence technology, and some of which cannot. Jobs and companies can be sliced and diced, and the process is already under way. Parts of the economies of developed countries are being fragmented, or Balkanised, with more and more people working freelance, carrying out individual tasks which are allocated to them by platforms and apps like Uber and TaskRabbit.

There are many words for this phenomenon: the gig economy, the networked economy, the sharing economy, the on-demand economy, the peer-to-peer economy, the platform economy, and the bottom-up economy.

Is this a way to escape the automation of jobs by machine intelligence? To break jobs down into as many component tasks as possible, and preserve for humans those tasks which they can do better than machines? Probably not, for at least two reasons. First, it is precarious, and secondly, the machines will eventually come for all the tasks.

Working for yourself can seem an appealing prospect if your current job is a poorly-paid round of repetitive and boring activities. There is freedom in choosing your own hours of work, and fitting them around essential features of your non-work life, like children and hangovers. There is freedom in choosing who you work with, and in not being subject to the arbitrary dictates of a vicious or incompetent boss, or the unfathomable rules and regulations of a Byzantine bureaucracy.

If you are lucky enough to be exceptionally talented, or skilled at a task which is in high demand, then you really can choose how and when you work. But freelancing can have its downsides too. Many freelancers find they have simply traded an unreasonable boss for unreasonable clients, and feel unable to turn down any work for fear that it will be the last commission they ever get. Many freelancers find that in hindsight, the reassurance of a steady income goes a long way to compensate for the 9 to 5 routine of the salaried employee.

Whether or not the new forms of freelancing opened up by Uber, Lyft, TaskRabbit, Handy and so on are precarious is a matter of debate, especially in their birthplace, San Francisco. Are the people hired out by these organisations "micro-entrepreneurs" or "instaserfs" - members of a new "precariat", forced to compete against each other on price for low-end work with no benefits? Are they operating in a network economy or an exploitation economy? Is the sharing economy actually a selfish economy?

Whichever side of this debate you come down on, the gig economy is a significant development. Gallup's most recent poll, in 2018, found that 29% of US workers were primarily employed in the gig economy, and adding those who worked partly as employees and partly in the gig economy took the number to 36%.[245]

Gallup distinguished between two types of gig employees: independent gig workers have autonomy and control. Contingent gig workers don't. Unsurprisingly, the former earned more money and were more content.

But our concern here is not whether the gig economy is a fair one. It is whether the gig economy can prevent the automation of jobs by machine intelligence leading to widespread unemployment. The answer to that is surely "no": as time goes by, however finely we slice and dice jobs into tasks, more and more of those tasks are vulnerable to automation by machine intelligence as it improves its capabilities at an exponential rate.

CHAPTER 18

SELF-DRIVING VEHICLES

At the time of writing, the consensus among people who think about the future of work seems to be that technological unemployment will not happen – at least not for many decades. Self-driving vehicles may turn out to be the canary in the coalmine – the technology which shifts this consensus, and persuades a majority that technological unemployment is a serious possibility within a generation or so, and that policies should be developed to cope with it. In chapter 6, we reviewed the history of self-driving cars, and their capabilities today. In chapter 10, we discussed when they might be ready for primetime, and I argued that the middle of the 2020s was a reasonable estimate. In this chapter, we will consider how long it will take for them to replace human-driven cars in various locations, and their effects on employment.

Driving jobs

There are nearly 4.5 million professional drivers in the USA: 3.5 million truck drivers,[246] 650,000 bus drivers,[247] and 230,000 taxi drivers.[248] How many of these jobs will be lost to machines, and when?

Once it is technically and economically feasible to replace human drivers with machines, it is a very short step to being economically necessary. Drivers account for a third to a half of the operating cost of the vehicles they drive. You can't escape the invisible hand of economics for long. In a free market, once one firm replaces its drivers the rest will have to follow suit, or go out of business. Trade unions and sympathetic governments will try to stop the process in some jurisdictions. They may succeed for a while, but only by rendering their industry uneconomic, and burdening their customers with unnecessary costs which will damage them in turn.

In July 2017, India's transport minister said "we won't allow driverless cars in India. I am very clear on this. In a country where you have unemployment, you can't have a technology that ends up taking people's jobs." Opposition politicians rolled their eyes and predicted the policy would be swiftly reversed, and even the minister himself gave himself an escape route: "Maybe some years down the line we won't be able to ignore it, but as of now, we shouldn't allow it."[249]

In fact the process has already started. In the Yandicoogina and Nammuldi mines in Pilbara, Western Australia, transport operations are now entirely automated, supervised from a centre in Perth, which is 1,200 miles away.[250] Mining giant Rio Tinto was prompted to take this initiative by economics: the decade-long mining boom caused by China's enormous appetite for raw materials. Drivers earned large salaries in the hazardous and inhospitable environments of these remote mines, which made the investment case for full automation irresistible.[251] The economics are going in the same direction everywhere – fast.

There are also safety considerations, and the bigger the vehicle, the more serious the concerns. Articulated lorries are driven by professional drivers who are well-trained, and who get lots of practice. Their backgrounds are checked, and their working hours and conditions are regulated. They cause fewer accidents per mile driven than the cars owned by the rest of us. But because they are much heavier, when they are involved in accidents they cause much more damage to life and property. Once machines can do this job, it is inconceivable that we will continue to allow humans to do it.

Not just driving

It is often argued that self-driving technology will not make human drivers redundant because driving is not the whole of the job. Drivers are often responsible for loading and unloading their vehicles. They also have to deal with the myriad surprises which are thrown at them by life, which is an untidy business at best. If a consignment of barbed wire falls off the back of a truck in front, they will get out and help.

Consider the process of delivering a consignment from a warehouse to a supermarket or other large retail outlet. Amazon's fleet of Kiva robots show that warehouses are well on the way towards automation. The unloading bays at the retail end are also standardised for efficiency: a system which automates the entire unloading process from a truck into the retailer's receiving area is technically feasible today, and with the exponential improvement in robotics and AI, it won't be long before it is economically feasible as well.

As we have seen, robots are becoming increasingly flexible, nimble and adaptable. They can also increasingly be remotely operated. Most of the situations a driver could deal with on the open road will soon be within the capabilities of a robot which does not need sleep, food or salary. On the rare occasion when human intervention is needed, the gig economy can probably furnish one quickly enough.

Skeptics about technological unemployment point out that planes have been flying by wire for decades, with human pilots in control for only around three minutes of an average commercial flight. We have yet to dispense with the services of human pilots.

However, a 747 is very different to a truck travelling down Highway 66. A truck is an expensive vehicle, and capable of inflicting severe damage, but commercial planes are on a different scale: they cost many millions of dollars each, and their potential to cause harm was graphically and tragically demonstrated in New York in 2001. Furthermore, those three minutes of human control are in part due to the difficulty of resolving the edge cases we discussed before. In road vehicles if not in planes, we are well on the way towards resolving those.

The roll-out

In chapter 10, I argued that self-driving vehicles will be ready for widespread use – at least in certain geographically restricted areas – by the mid-2020s. Once they do start to appear outside the current very limited pilots, how long will they take to replace significant numbers of human drivers, and how will the transition be phased geographically and sectorally?

No-one knows for sure, but the initial use cases for Autos will probably be delivery vehicles and taxis in cities. This is because early Autos will be expensive: they need massive on-board compute power, and lots of sensors, including (most people think) Lidars. Although few consumers are likely to spend the necessary money, commercial vehicle operators will have to.

Cities are likely to lead the way because that is where the demand is: more people, more traffic, more money. A well-known pair of photos show the same stretch of Broadway in New York City on Easter Day in 1900 and 1915. In 1900, there was one motor car, and every other vehicle was horse-drawn. In 1915, every vehicle except one was a motor car.[252] So it took 15 years to swap horses for internal combustion engines in New York City. The transition to self-driving cars is different, but maybe the timing will be similar for taxis, buses, and for vehicles delivering parcels within major cities.

Given the commercial incentive and the safety issues, long-haul trucking may transition in a similar time period. So if we allow five years to reach prime time, and fifteen years for general transition from human-driven cars to self-driving vehicles, professional drivers may be getting scarce as we approach the year 2040.

In the case of consumer-owned vehicles, Elon Musk has made the interesting observation that the first owners of self-driving cars could use their vehicle to travel to work (or school) in the morning, and then offer it via an app like Uber to other people who need transportation. Thus the early adopters of self-driving cars could offset the purchase price – perhaps to zero. That would boost adoption significantly.

Frictions

Of course, just because a product becomes available, that doesn't mean it will be bought, still less that it will comprehensively replace the existing population of products that it is designed to supplant. The rate at which that happens, if it happens at all, depends on a host of factors including regulation, price, design, service support, promotion and PR.

The United States Congress has been trying to provide a clear federal regulatory structure for Autos since 2017, and failing. The fear of being overtaken by Chinese Auto manufacturers has not yet been sufficient to drive an agreement through.[253] In theory, unco-operative regulators could slow or even stop the arrival of self-driving cars, and there will be powerful lobbies pressing for this. But they can only succeed if all regulators everywhere agree, and work together, and that will not happen – even within the US, never mind globally.

Increasingly, the regulatory authorities in many countries are keen to encourage self-driving technology, partly to save lives, and partly because they realise there is a commercial advantage to be gained.

Other affected industries

Automotive cover represents 30% of the insurance industry, so a shift to self-driving cars will have a major impact on that industry. The most obvious effect should be a sharp reduction in pay-outs because there will be far fewer accidents. This in turn should mean far lower premiums: bad news for the insurance companies, good news for the rest of us.

Who will take out the insurance policy? When humans drive cars we blame them for any accidents, so they pay for the insurance. When machines drive, does the buck stop with the human owner of the vehicle, the vendor of the self-driving AI system, or the programmer who wrote its code? If the insured parties are Google and a handful of massive competitors, then the negotiating position of the insurance companies will deteriorate sharply from the present situation where they are "negotiating" with you and me.

Warren Buffet ascribes some of his enormous success as the world's best-known investor to his decision to avoid areas he does not understand, including industries based on IT. He has massive holdings in the insurance industry. Unfortunately for him, software is "eating the world",[254] and a large chunk of the insurance industry is about to be engulfed in rapid technological change. Buffet acknowledges that when self-driving cars are established, the insurance industry will look very different, almost certainly with fewer and smaller players.[255] It is very hard to say which of today's players will be the winners and losers.

The law of unintended consequences means that we cannot say how the insurance risks will change. Let's hope this never happens, but what if a bug – or a hacker – caused every vehicle in a particular city to turn left suddenly, all at the same time? How does an insurance company estimate the probability of such an event, and price it? Important issues like this might slow down the introduction of self-driving cars, but they will not stop it. They are capable of resolution, just as we resolved questions about who would build the roads and who would have the right of way in different traffic situations in the decades after the first cars appeared.

People working in insurance companies will certainly not be the only ones affected by the move to self-driving cars. Machines will presumably be programmed not to violate local parking restrictions – and they will have no need to do so. That will remove a significant source of income from local authorities: parking charges generate well over $300m a year for the city government of Los Angeles.[256]

Automotive repair shops will still be needed, but their business will shrink as it becomes restricted to maintenance and repairs necessitated by age rather than accidents. Happily, something similar can be said of doctors and nurses.

Road rage against the machine

If and when it happens, the automation of driving will have a major impact on the overall job market. One in fifteen American workers is a truck driver,[257] and truck and delivery driver is the most commonly named occupation in 29 US states.[258] These drivers and their unions are unlikely to surrender their jobs without a struggle, and there may even be violence – call it road rage against the machine. Perhaps it will only be violence against property, with the cameras and sensors of self-driving vehicles being spray-painted over. Perhaps there will be violence against the owners of fleets which adopt self-driving vehicles.

Much depends on how governments handle the transition. At the moment, very few of them are even considering the issue. In places like Pittsburgh, which has become a centre for self-driving technology, they welcome the new jobs that have been created. They are great jobs: highly-paid employment for PhDs, plus a few jobs for the chaperones who sit in the test cars and take over driving duties every now and then. The people of Pittsburgh can see that these will not replace all the driving jobs that will be lost, but the politicians have no answers.[259]

Of course it won't only be truck drivers. In 2014 the medallion that gives you the right to drive a yellow taxi in New York City sold for $1.3m. Drivers and investors took out huge loans to buy them, but their price has crashed as Uber, Lyft and others have entered the market. Many people who thought they had achieved financial independence now find themselves bankrupt, or working ever harder to meet interest payments on the loans.

This is just a foretaste of what is to come. When self-driving cars become a common sight on roads around the world they may be like the canary in the coal mine which gave miners a bit of advance warning that they were in great danger. Self-driving vehicles may have the effect of alerting everyone else to the prospect of widespread technological unemployment.

CHAPTER 19

WHO'S NEXT?

The Frey and Osborne paper

Carl Benedikt Frey and Michael Osborne are the directors of the Oxford Martin Programme on Technology and Employment.[260] In 2013 they produced a report entitled "The future of employment: how susceptible are jobs to computerisation?" It has been widely quoted, and its approach to analysing US job data has since been used by others to analyse job data from Europe and Japan. You can read more about this report in the appendix. It is the report which got the ball rolling on the current round of discussion about technological automation.

The report analysed 2010 US Department of Labour data for 702 jobs, and in a curious blend of precision and vagueness, concluded that "47% of total US employment is in the high risk category, meaning that associated occupations are potentially automatable over some unspecified number of years, perhaps a decade or two." 19% of the jobs were found to be at medium risk and 33% at low risk. Studies which have extended these findings to other territories have yielded broadly similar results.

The report suggested that the automation would come in two waves. "In the first wave, we find that most workers in transportation and logistics occupations, together with the bulk of office and administrative support workers, and labour in production occupations, are likely to be substituted by [machines]."[261] In other words, lower income jobs will be the first to be automated. It makes intuitive sense that lower income jobs would be less cognitively demanding, and hence easier to automate.

Lower income service jobs

Arriving at JFK airport is a very mixed blessing. It is always exciting to visit New York City, but immigration control at JFK often seems to be run by sadists offering a main dish of rudeness and a side order of incompetence. How nice it would be to disembark from a plane after a long-haul flight and wait in comfort (maybe even seated!) while robotic baggage handlers fetch your luggage, and unintrusive scanning drones quickly and efficiently check your face against the documents you have transmitted. While JFK is reported to be getting a $10bn overhaul, airports in Asia are already investing and competing to provide much better experiences, since that is where the growth markets are.[262] In the medium term, we can expect to be processed by many fewer humans at airports, and we will probably not regret it.

As you read this, there are people all over the world inspecting pipes and cables. This is lonely, often boring, and occasionally dangerous work, but it has to be done. Not by humans for very much longer, though. Smart drones are already taking over in some industries, and much of the rest will follow.[263] In a decade or two, the Wichita lineman won't be still on the line.

Retail

Automation is not new to the retail sector, as we noted in chapter 16. McKinsey estimated in May 2019 that currently available automation technologies could enable a typical grocery store to reduce its man hours by up to 65%. The technologies included self-checkout terminals, electronic shelf labels, shelf-scanning robots, and partially automated store room unloading.[264]

Amazon placed itself at the cutting edge with its Go store concept, where there is not only no human at the checkout – there is no checkout. You pick the product from the shelf, put it in your bag, and leave the store. Cameras identify your purchases, and AI systems take care of the billing. By March 2020 Amazon had opened 27 of these stores, in four US cities, and had announced plans for 3,000 stores by 2021.[265]

Food service

Automation is not new in food service, either. Back in 1941, the Automat chain served half a million American customers a day, dispensing macaroni cheese, baked beans and creamed spinach through cubby-holes with glass doors.[266] The chain declined in the 1970s with the rise of fast-food restaurants serving better-tasting food, such as Burger King and, of course, McDonalds. But these chains themselves are now discovering the economic appeal of automation.

Reservation management, order-taking, meal preparation, and table service are all areas where the chains are experimenting and investing. McDonalds was reported to be spending $1bn in 2019 on automated kiosks.[267] In the same year, Wendys launched a plan to install self-ordering kiosks at 1,000 outlets, at a cost of $15,000 each, with a projected payback period of two years thanks to reduced labour costs and increased sales.[268]

There has been heated political debate about whether these and similar initiatives are prompted by increases in minimum wage levels, but the truth is that AI systems and robots are gradually becoming more cost-efficient than humans.

Customer preference

Cost savings are not the only reason for this kind of automation. In many situations, humans prefer to transact with machines rather than other humans. It can be less time-consuming, and it requires less effort. It can also make a service available for longer hours, perhaps 24 hours a day. Automated Teller Machines (ATMs) are the classic example. Another are the automated passport control systems now installed at many airports, which many people opt to use in preference to the manned channels.

A report published in 2015 by Forrester, a technology and market research company, claimed that 75% of procurement professionals and other people buying on behalf of businesses (i.e., B2B buyers) prefer to use e-commerce and buy online rather than deal with a human sales representative. Once the buyers have decided what they want, the percentage rises to 93%.[269] Forrester pointed out that many vendors were ignoring this fact, and obliging customers to speak to a human. This is no doubt partly at least because human sales people are currently much better able to up-sell the buyer, but this is also one of the reasons why buyers prefer e-commerce. Forrester argued that companies which wait too long to offer good e-commerce channels risk losing market share to more digitally-minded competitors.

Call centres

We are still at the very early stages of introducing artificial intelligence to call centres. For many of us, dealing with call centres is one of the least agreeable aspects of modern life. It normally involves a good deal of waiting around, listening to uninspiring hold music, followed by some profoundly unintelligent automated routing, and finally a conversation with a bored person the other side of the world who is reading from a script written by a sadist.

The two largest providers of contact centre services to global corporations are the US firm Concentrix and France's Teleperformance. In recent years they have seen the volume of calls handled by human call centre agents shrink as a percentage of their work, as chatbots handle more and more of customers' basic enquiries, and other types of digital interaction grow. But the absolute number of calls handled has been stable, and their employment numbers have grown modestly.

As the most basic types of customer interactions are automated, the human staff in contact centres can provide more sophisticated assistance – often with the aid of machines, which trawl through databanks to generate answers promptly, and which can suggest alternative directions to steer the conversation in real time.

Contact centres are a big business, employing six million people in the US, for instance, and well over a million people in the Philippines, which has overtaken India to become the leading provider in the developing world. A senior industry figure commented in 2020 that "most vendors offer clever natural language processing technology, but it is not yet real AI. Some exceptions are coming through, but the software is still mostly pre-programmed, using lookup tables and knowledge banks. [But] of course, the endgame - in the not too distant future - is that many aspects of even my job can be done pretty much by a machine."

Physical robots are often harder to develop and perfect than the software robots ("softbots" or just "bots") which are automating basic, repetitive cognitive tasks in offices and shops. But manual jobs are being and will be automated in agriculture, construction, manufacturing and distribution.

Manual work: agriculture

As we saw in chapter 16, agriculture was the poster child for mechanisation (mechanical automation) during the industrial revolution. It is not leading the way in cognitive automation, as the number of people involved is much smaller now: only a third of a million people work in UK agriculture today.[270] But automation is very much a feature of modern agriculture.

Tractors, the workhorses of 20th-century farming, are increasingly using technologies developed for self-driving cars. Major manufacturers like John Deere, Case, and New Holland are competing to drive down the cost of automation, and drive up the functionality. Companies like Rabbit champion the idea that swarms of small vehicles can replace the huge machines which are capable of dozens of acres of production an hour, and do it more efficiently, more cost-effectively, and more sustainably.

Spraying and weeding equipment scan rows of crops as they go, identifying individual plants, and applying fertiliser or herbicide only as required. A strawberry-picking machine produced by Octinion can pick 340kg of fruit a day, seven times as many as a human.[271] In June 2019, 10% of British cows were being milked by robots. [272]

Harper Adams University is a specialist provider of higher education for the agricultural sector in the UK, and in 2017 they managed to plant, tend, and harvest a "hands-free hectare" of barley using only autonomous vehicles and drones.[273] In 2019 they expanded the area under cultivation to 35 hectares.[274]

Manual work: factories and warehouses

Manufacturing accounts for over a third of China's GDP, and employs more than 100 million of its citizens. Historically, China's competitive strength in manufacturing has been its low wage costs, but this is changing fast: wages have grown at 12% a year on average since 2001, and Chinese manufacturers are embracing automation enthusiastically. China is now the world's largest market for industrial robots, and as we saw in chapter 11, its penetration levels (the number of robots per worker) are comparable to the USA's and Europe's, although well behind South Korea's.

Industrial robots are far from perfect, and manufacturers and industry observers alike have a track record of under-estimating the time required for significant progress. In 2011 the CEO of Foxconn, a $130bn-turnover Taiwanese manufacturer that is famous for making iPhones, declared a target of installing a million robots by 2014. The robots failed to perform as he hoped, and the actual installation rate has been much slower. But they are improving.[275]

Warehousing is a big business, employing over a million Americans. The automation of warehouses is accelerated by the retail industry's move from bricks and mortar to e-commerce. Global online retail sales are still modest, at 9% of total retail sales in 2018, but that is a 20% increase on the 2016 figure.[276] (The UK decided long ago that it prefers congestion to building new roads, so its consumers are enthusiastic adopters of online shopping, and its e-commerce penetration is double the global average, at 16% of total retail.)

The Covid-19 pandemic provided a major boost for ecommerce, with internet sales rising to 33% of total retail sales in the UK in May 2020. It is too soon to say how much of this shift will be reversed as the lockdown eases, and people work out what their new normal will look like.[277]

Amazon is one the biggest beneficiaries of this structural shift to online retail. To drive costs down, it is working hard to automate as much of its supply chain as possible. The company's Kiva robots now fetch items from shelves and bring them to work stations, where humans carry out the picking and packing tasks which are hard for robots thanks to Moravec's paradox. Every year Amazon holds a competition for developers of robotic pickers, with a $250,000 prize for the team which solves the problem. Alberto Rodriguez, a roboticist at MIT who has worked with Amazon on the problem, estimated in 2017 that it would take five years or so.[278]

Youtube has plenty of videos showing how far automation has advanced in warehouses, but people often point to Amazon's ravenous hiring of warehouse people to show that "it's a myth that automation destroys net job growth," to quote Dave Clark, a senior executive at operations there.[279] It will not always remain a myth. In 2019, Amazon had 100,000 robots in its US warehouses, and 200,000 globally,[280] but it is still hiring humans because it is growing fast, and the robots cannot yet do everything the humans can. Amazon's no-profits business model and its strong management means that it is eating the lunches of bricks and mortar retailers.

The professions

It is certainly not only manual jobs, or low-paid white-collar jobs that will be automated in the coming decades. The professions – jobs where people undergo extensive training in order to provide disinterested advice - are vulnerable too. In the rest of this chapter, we will look at trends in journalism, medicine and education. In the next chapter will follow the money, and review finance and the law.

Journalists

In 2014, the Los Angeles Times published a report about an earthquake three minutes after it happened. This feat was possible because a staffer had developed a bot – Quakebot – to write automated articles based on data generated by the US Geological Survey.[281]

Narrative Science, a company established in Chicago in 2010, is one of the leading companies providing Natural Language Generation (NLG) tools. By the mid-2010s, its StatsMonkey system was writing thousands of articles every day on finance and sports for media outlets like Forbes and Associated Press (AP).[282] Most readers could not identify which articles were written by StatsMonkey and which by human journalists, and StatsMonkey was much faster. Narrative Science has since moved on to the more lucrative terrain of helping corporates to produce internal reports with a system called Quill, but many media outlets have developed their own in-house equivalents.

Systems like StatsMonkey and Quill start with data – graphs, tables and spreadsheets. They analyse these to extract particular facts which could form the basis of a narrative. They generate a plan, or narrative for the article, and finally they craft sentences using natural language generation software. They can only produce articles where highly structured data is available as an input, such as video of a football match, or spreadsheet data from a company's annual return. They cannot write articles with flair, imagination, or in-depth analysis.

StatsMonkey and its peers have not rendered thousands of journalists redundant. Instead it has sharply increased the number of niche articles being written. Newspaper revenues have declined sharply since the turn of the century, as classified ads for jobs, houses and cars migrated online. News services like AP responded by increasing the daily quota of articles for each journalist, cutting back the number employed, and reducing the number of articles produced on the quarterly earnings reports of particular companies, for instance. NLG tools have enabled them to reverse those reductions. AP now produces articles on the quarterly reports of medium-sized companies that it gave up covering years ago.

Kristian Hammond, founder of Narrative Science, forecast in 2014 that in a decade, 90% of all newspaper articles would be written by AIs. However, he argued that the number of journalists would remain stable, while the volume of articles increased sharply. Eventually, articles could become tailored for particular audiences, and ultimately for each of us individually. For instance, an announcement by a research organisation that inflating your car tyre correctly could reduce your spend on petrol by 7% could be tailored – perhaps with the help of your Digital Personal Assistant - to take into account your particular car, the number of miles you drive each week, and even your style of driving. (Although of course by then you will perhaps not do the driving yourself anyway.)

AI systems are treading on the toes of TV presenters as well as print and online journalists. In 2015, Shanghai's Dragon TV introduced Xiaoice, (pronounced Shao-ice) an AI weather presenter with a remarkably life-like voice,[283] based on the Mandarin version of Microsoft's Cortana digital assistant software. Audience feedback was positive.[284] In 2020, Microsoft claimed that Xiaoice had become the world's most popular chatbot, with 600m regular users, and span it out into a separate company.[285]

Icebergs

The development of journalism described above is an example of the iceberg phenomenon. AI systems have expanded the output of the sector dramatically, but they have not replaced human journalists. Yet. Journalists were like a group of people standing nervously on a body of ice, thinking they were only separated from freezing water by a thin layer. Then they discovering that in fact they were standing on an iceberg, with a huge mass of previously unknown solid ice beneath their feet. AI systems have increased media output enormously, making it possible to cover relatively minor companies and sports teams.

We will see this phenomenon in the other professions too. In the short and medium term, machine automation of white-collar jobs opens up vast new areas of work that can be undertaken, and doesn't throw the incumbent humans out of work. Humans are still needed to train the system at the start of a large new assignment, and to execute more sophisticated tasks.

Economist James Bessen called this the automation paradox.[286] But he also observed, "[It] could change decades into the future, as new generations of software powered by artificial intelligence becomes ever more capable of advanced tasks." We don't know how long the iceberg phenomenon will last.

Other writers

Not everyone who spends their working days crafting crisp sentences is a journalist. They might be PR professionals, or online marketers, for instance. A company called Persado claims that marketing emails drafted by its AI have a 75% better response rate than emails written by human copywriters.[287] Citibank and American Express are customers as well as investors.

In January 2016 a researcher at the University of Massachusetts announced an AI which can write convincing political speeches for either of the two main US political parties. The system learned its craft by ingesting and analysing 50,000 sentences from Congressional debates.[288] In 2018, IBM unveiled Project

Debater, an AI system capable of engaging in real-time debates with human experts. This was IBM's third great AI challenge, after its chess-playing Deep Blue in 1996, and its Jeopardy-playing Watson in 2011.[289]

Doctors

Doctors are a scarce resource. Only bright and dedicated people are admitted to the relevant university and post-graduate courses, and these courses demand many years of hard study. Hospitals and local surgeries are organised to maximise the availability of this resource, but some critics argue that they are organised for the benefit of the doctors rather than the patients. In 2015, senior doctor and medical researcher Eric Topol published a book called "The Patient Will See You Now", which he argues should become the mantra for the profession, replacing the current one, which he says is "the doctor will see you now".

Suggesting acerbically that the initials MD stand for Medical Deity, Topol accuses many doctors of being arrogant and paternalistic towards their patients, assuming they are unable to understand the detailed information regarding diagnoses, and withholding information from patients so as not to upset them. He believes that the digital revolution will start to overturn this unsatisfactory state of affairs, as it will place cheap and effective diagnostic tools in the hands of patients.

Diagnosis

In chapter 3 we met Geoff Hinton, often described as the father of deep learning. In 2016, he said "If you work as a radiologist, you're like the coyote that's already over the edge of the cliff, but hasn't yet looked down so doesn't realise there's no ground underneath him. People should stop training radiologists now. It's just completely obvious that within 5 years, deep learning is going to do better than radiologists." Andrew Ng, another prominent AI researcher said that a "highly-trained and specialised radiologist may now be in greater danger of being replaced by a machine than his own executive assistant."[290]

In the following years, a string of papers and reports described and demonstrated AI systems which achieved human-level accuracy in diagnosing a range of ailments from various types of scans. But Hinton and Ng's enthusiasm proved premature. Partly because sometimes the AI systems were picking up clues which had nothing to do with the medical situation. For example, in one study, the scans of people with cancer happened to have a ruler in the image, whereas the scans of people without cancer did not. The AI system detected this correlation and achieved a high degree of accuracy based on an irrelevant correlation.

As a result, the number of radiologists who have been replaced by AI systems to date is "approximately zero".[291] Radiologists do more than just staring at images, and for some years to come, they will be assisted by machines, not replaced by them.[292]

Remote diagnosis

AI systems will facilitate an enormous improvement in diagnosis, known as telemedicine. The first step in this is diagnosis by video call. A British startup called Babylon charges customers £5 a month for phone (and videophone) access to a dedicated team of doctors. Before a doctor comes on the line, the patient is triaged by a machine.[293] As AI improves, the role of the human doctor in this process will continuously be reduced.

The next logical step is continuous monitoring of our health indicators by devices we carry all the time. Smartphones are increasingly able to gather medical data about us, and perform basic analysis. By attaching cheap adapters to their phones, patients can quickly take their blood pressure, sample their blood glucose, and even perform an electrocardiagram. Your breath can be sampled and digitised, and used to detect cancer, or potential heart problems. Your camera's phone can help screen for skin cancers. Its microphone can record your voice, and that data can help gauge your mood, or diagnose Parkinson's disease or schizophrenia.

All this data can be analysed to a certain level within the phone itself, and in many cases that will suffice to provide an effective diagnosis. If symptoms persist, or if the diagnosis is unclear or unconvincing, the data can be uploaded into the cloud, i.e., to server farms run by companies like Amazon and Google. The heart of diagnosis is pattern recognition. When sophisticated algorithms compare and contrast a set of symptoms with data from millions or even billions of other patients, the quality of diagnosis can surpass what any single human doctor could offer.

The medical iceberg

As with journalists, the improvement in diagnostic machines does not mean that doctors will become unemployed any time soon. There is vast reservoir of unmet healthcare needs – needs which automation and machine intelligence can help to satisfy. Formed in Mumbai in 1996, Thyrocare Technologies is the world's largest thyroid testing laboratory. Its founder, Dr A Velumani, had the insight that 90% of people who could benefit from diagnostic tests were not receiving them because they were too expensive, so the tests were restricted to those already manifesting symptoms of disease. He established Thyrocare to address this latent demand, and by 2015 it was processing 40,000 samples a day.[294]

Operations

You might think that the hands-on physical and frankly messy business of surgical operation will remain the preserve of humans rather than machines for the foreseeable future. Probably not. One of the most highly skilled professionals in the emergency suite is the anaesthetist, and Johnson & Johnson has an automated version called Sedasys. Despite fierce opposition from the profession, Johnson & Johnson secured FDA approval for Sedasys to provide anaesthesia in less challenging procedures like colonoscopies. It carried out thousands of operations in Canada and the USA.[295]

In March 2016, Johnson & Johnson announced that it was exiting the Sedasys business due to sluggish sales, despite the machine costing $150 per operation whereas a human anaesthetist costs $2,000.[296] This will certainly not be the last setback in the progress of the machines, but in the long run the economic facts will prevail - although perhaps more slowly in industries which are not subject to market economics.

In May 2016 an academic paper announced that a robotic surgeon had out-performed human peers. The Smart Tissue Autonomous Robot (STAR) operated on pig tissue and did the job better, although four times more slowly, than humans operating alone - and also better than humans aided by the semi-robotic Da Vinci system.[297]

In September 2017, Hong Kong's newspaper of record reported that the world's first successful automated dental implant surgery had fitted two new teeth into a woman's mouth. The patient looks terrified in the article's picture, but it was her first time as well as the world's.[298]

A study published in 2017 in the Lancet medical journal found that most of the 20,000 British men who had prostate surgery in the first half of this decade preferred to have their operations in clinics which used robots. This preference was strong enough to force some of the clinics without robots to close.[299]

Education

Teachers are the active ingredient in education – at school level, anyway. Studies have shown repeatedly that the quality of teaching makes an enormous difference to how well a student performs at school and afterwards. But schools cannot afford enough of them, governments burden them with bureaucracy, and most countries under-value them.

What happens when the learning of every pupil is monitored minutely by artificial intelligence? When every question she asks and every sentence she writes is tracked and analysed, and appropriate feedback is provided instantly? Teachers will play the role of coach instead of instructor, but as with the other professions, their scope for contribution will shrink.

The beachhead for AI in education is marking, also known as grading. This is the bane of many teachers' lives, and they will welcome an assistant which can relieve them of the duty. A company called Gradescope marks the work of 55,000 students in 100 US universities, marking simple, multiple-choice types of test. It raised $2.6 million in April 2016 to develop its product into complex questions and essays.[300] Large corporates like Pearson and Elsevier which provide education services are moving in the same direction.

Towards the end of 2015, 300 students at Georgia Institute of Technology were unwitting guinea pigs in an experiment to see whether they would notice that one of their nine teaching assistants was a robot. Only ever in contact via email, they would ask questions like "Can I revise my submission to the last assignment?" and receive answers back like "Unfortunately there is not a way to edit submitted feedback." None of the students noticed that Jill Watson, named after the IBM Watson system "she" ran on, was in fact an AI.[301]

CHAPTER 20

FOLLOW THE MONEY

We saw in chapter 18 that the jobs of professional drivers may be automated in the medium term, and in chapter 19 we added warehouse workers and call centre staff. Professionals like journalists, medics and teachers may take a while longer to feel the chill wind of automation, partly because of the iceberg phenomenon. What about the more monied professions?

The legal profession, and parts of the financial services industry, are sometimes accused of being conspiracies against lay people. They are protected occupations, with demanding entry requirements and restrictions on the number of trainees who can join each year. They command prestige and high salaries, but that may not last forever.

Lawyers

Whatever Hollywood thinks, most lawyers do not spend their days pitting their razor-sharp wits against equally talented adversaries in front of magisterial judges, eliciting gasps of admiration from around the courtroom as they produce the winning argument with a flourish. Most of the time they are reading through piles of very dry material, looking for the thread of evidence which will convict a fraudster, or the poorly drafted phrase which could undermine the purpose of a contract.

Discovery

Many lawyers get a lot of their on-the-job training through the "discovery" process. Known as "disclosure" in the UK, this is a pre-trial process in civil law in which both sides must make available all documents which may affect the outcome of the case. An analogous process takes place in the "due diligence" phase of a corporate merger or acquisition (M&A), in which teams of junior lawyers (and accountants) spend weeks locked away in data rooms, reading through material which can run into millions of documents, looking for something which would clinch the case, or, in the case of M&A work, provide a reason to terminate or renegotiate the deal.

Looking for a needle of fact in a haystack of paper is work more suited to a machine than a man. And although lawyering is a very conservative profession, it finds itself at the forefront of cognitive automation. RAVN Systems is the British company behind an AI system called Ace, which reads and analyses large sets of unstructured, un-sorted data. It produces summaries of the data, and highlights the documents and passages of most interest according to the pre-set criteria.[302] When one of the UK's largest law firms started working with Ace in 2016, it was regarded as pioneering, experimental, and somewhat risky. Two years later, that law firm was promoting its own services to potential new clients on the basis that it knew best how to exploit the advantages of RAVN Ace. Two years is a very short time for anything at all to happen in the legal industry!

Typically, a new client's data will present a new set of challenges. It usually takes a few days to train the system how to read the data, which currently involves human intervention. Once the training is complete, though, the work proceeds without human involvement, and the system will finish the work much faster than human lawyers could. This means that law firms are having to work out new ways of billing their clients: the old system of hourly rates is under challenge.

The legal iceberg

Forward-thinking lawyers are excited about the arrival of this sort of automation. They have realised it is another example of the iceberg phenomenon. Rather than fearing that cognitive automation will destroy the jobs of junior lawyers, making it impossible for young people to learn the profession, they believe it will increase the amount of cases that can be handled.

Imagine that a large supermarket chain wants to know the implication of making a small change to the employment contracts of all its in-store employees – tens of thousands of people. Previously, its employment law firm would have said that this task could not be undertaken cost-effectively with any degree of rigour. RAVN Ace and systems like it make this kind of work possible, opening up whole new avenues of work for law firms.

As Greg Wildisen, MD of Neota Logic, a firm providing an AI platform for lawyers, puts it, "So many legal questions go 'un-lawyered' today that there is enormous scope to better align legal resources through technology rather than fear losing jobs."[303]

But as RAVN Ace and its successors improve – at an exponential rate, of course – they will be able to take on more and more of the sophisticated and demanding aspects of the lawyers' work. No-one can be absolutely sure yet whether this process will hit a wall at some point, leaving plenty of work for human lawyers, or whether it will continue to the point where there are very few jobs left for humans.

Forms

Another fairly basic form of legal work is the completion of boilerplate (standard) forms to establish companies, initiate a divorce, register a trademark, request a patent and so on. A company called LegalZoom was established in 2001 to provide these services online, and increasingly, to automate them. LegalZoom now claims to be the best-known law brand in the US,[304] and in 2014 the private equity firm Permira paid $200m to become its largest shareholder. Another company, Fair Document, helps clients complete forms for less than $1,000, one-fifth the amount it would have previously cost.[305]

More sophisticated lawyering

At the other end of the spectrum from the "grunt" work of discovery and filling out legal forms, one of the most sophisticated and important jobs that senior and successful lawyers are asked to undertake is to estimate the likelihood of a case winning. The advice is vital, as it will determine whether large amounts of money are spent. A team at Michigan State University developed an AI system in 2014 that analysed 7,700 US Supreme Court cases. It predicted the verdicts correctly 71% of the time.[306]

Another job for experienced lawyers in common law jurisdictions such as the US and the UK is identifying which precedent cases to deploy in support of litigation. A system called Judicata uses machine learning to find the relevant cases using purely statistical methods, with no human intervention.[307]

In September 2016, the eastern Chinese province of Jiangsu started using "legal robots" to review cases and advise on sentencing. In the following year they reviewed 15,000 cases, many of them traffic violations. They detected errors in more than half the cases.[308]

Experienced patent lawyers are highly skilled and articulate people, but much of the work involved in securing a patent is routine and could perhaps be automated. In November 2015, I took part in a debate at the IMAX cinema in London's Science Museum. The motion was "This House believes that within 25 years, a patent will be applied for and granted without human intervention." Patent lawyers comprised a good part of the audience, and although the motion was vigorously opposed by two senior patent lawyers, the motion was passed. Not exactly turkeys voting for Christmas, but certainly food for thought.

Financial services

The finance sector is an obvious target for machine intelligence, with high-value and high-priced services based on vast amounts of data. "Fintech" is one of the hottest areas for venture capital investment.

As well as looking to invest in other companies using AI, venture capitalists are using AI to enhance their own business processes. Swedish VC firm EQT Ventures developed a system called Motherbrain to source potential deals. Since its launch in 2016, the system has scanned 10 million startups, and the executives claim that four of the firm's 40 portfolio companies were identified entirely by the system.[309]

The provision of advice to other kinds of investors is aided by machines too. A system called SigFig incorporates a client's risk appetite and investment style into its algorithms' analysis of low-cost opportunities and recommendations.[310] Similar so-called "robo-adviser" services are available from Betterment, Wealthfront and Vanguard.[311]

Banking

Banking, especially retail banking, has traditionally been a conservative, slow-moving industry, but bankers are spending considerable amounts of time and energy on working out where the most powerful disruption to their business models will come from, and whether they can do the disrupting themselves rather than be its victims.

In September 2017, the CEO of Deutsche Bank made headlines when he told an audience that "in our bank we have people doing work like robots. Tomorrow we will have robots behaving like people. It is going to happen. The sad truth for the banking industry is, we won't need as many people as today."[312]

The CEO of Mashreq Bank in Dubai agreed. "By using artificial intelligence, employment at banks will shrink over time," he said in 2017, adding that his own bank would shed 10% of its 4,000 people in the next 12 months. The average cost of its junior employees was Dh250,000, and the cost of replacing them with AI was a one-time investment of Dh30,000, he added. The kinds of services that can now be done with AI include issuing new credit cards, depositing checks and opening new accounts.[313]

The CEO of UBS said that his bank could reduce headcount from nearly 100,000 to around 70,000 over the next few years. It does not sound very comforting to hear that the bank "can have 30 percent [fewer people], but the jobs are going to be much more interesting jobs, where the human content is crucial to the delivery of the service."[314]

Trading and execution

Financial traders employ primitive forms of AI at the moment. In 2019 it was estimated that 14% of the world's 10,000 hedge funds (which collectively manage

$3 trillion for their clients) use computer models to make the majority of their trades.[315] But they are using traditional statistical methods rather than AI which learns and evolves. This is changing. In 2015, Bridge-water, the world's largest hedge fund, hired David Fer-rucci away from IBM, where he had project managed the development of the version of Watson which beat Ken Jennings at Jeopardy.[316] January 2016 saw the in-augural trades of Aidyia, a hedge fund based in Hong Kong whose chief scientist is Ben Goertzel, a promi-nent researcher in artificial general intelligence.[317]

The AHL hedge fund started using AI seriously in trading in 2014, and within a year it was generating half of the division's profits. With $5 billion under man-agement, AHL is part of the Man Group, the world's largest publicly-traded hedge fund, managing $96 bil-lion.[318] Nick Granger, the executive who persuaded the company to trust the AI became AHL's CEO. "The idea that the humans will just disappear … is just not right. It's just that they move to different tasks, to higher val-ue-added tasks. We need smarter humans than we did." This begs the question of what happens to the exist-ing staff when the bank has hired lots of these smart-er humans. Luke Ellis, the CEO of the parent Man Group is less sanguine: "My hope is always that there will be parts that humans do that AI doesn't do, [but] I wouldn't bet my life on it."[319]

Other experts agree. "The human mind has not become any better than it was 100 years ago, and it's very hard for someone using traditional methods to juggle all the information of the global economy in their head," says David Siegel of Two Sigma, another hedge fund which uses AI. "The time will come that no human investment manager will be able to beat the computer."[320]

Jeff Tarrant of investment firm Protégé Partners thinks that the technology is in an early phase of adoption, and "there is going to be mass unemployment in asset management in the next several years."[321]

"Algo trading" has many critics in financial circles, who point out that they chase spurious correlations (such as the fact that divorce proceedings in Maine have consistently tracked sales of margarine), and that they can move markets in ways that are impossible to follow and are potentially dangerous. But in a financial world become so complex that mere humans can no longer follow it, they may not only be inevitable, but also necessary. David Siegel says "People talk about how robots will destroy the world, but I think robots will save it."

In July 2017, JP Morgan, the world's biggest investment bank, announced that a trial in Europe of a deep reinforcement learning system for the execution of trades had been successful, and the system, known as LOXM, would be rolled out to the US and Asia. The execution of trades involves no selection of which equities to buy, but decisions about when to carry out the trades and in what volumes can significantly affect the price achieved, and were previously considered a skilled task for humans. JP Morgan claimed the move would give it a two-year advantage over rivals, which would cost them many millions to erase.[322]

Top analysts are among the highest-paid people in investment banks. They are targeted by a number of fintech companies like Kensho, which sorts through thousands of data sets to produce reports in minutes which would take skilled humans days. Kensho's founder, Daniel Nadler, said in 2016 that between a third and a half of finance employees would be redundant within a decade.[323]

Compliance

One of the drivers of AI use by financial services firms is the ever-growing and increasingly complex web of compliance requirements imposed by governments and regulators. Systems like IPSoft's Amelia help insurance firms and other financial services companies to navigate this web and make sure the forms and procedures used by staff are up-to-date.[324]

Global banks are regularly fined hundreds of millions of dollars for carrying out illegal or sanctioned trades. Standard Chartered's regulatory costs rose 44% in 2015 to $447m as it was obliged to hire thousands of additional staff to deal with compliance requirements. In March 2016 it announced a major investment in AI systems to oversee its traders' behaviour, and to match their activities against regulatory norms.[325]

CHAPTER 21

YES, IT'S DIFFERENT THIS TIME

Reprise

In part one of this book we considered the history of artificial intelligence, and the state of the art today. In part two we peered a little into the future, and we saw how dramatic the impact of exponential growth can be. (No apologies for repeating that point: it is critical.)

In chapter 16 we saw that previous rounds of automation in the industrial revolution have not caused widespread unemployment among humans in the long-term – although the long-term is fairly long in this context: the Engels pause lasted at least a generation (a quarter-century). And the horses were not so lucky.

In chapters 18, 19 and 20, we discussed how various occupations could be automated, starting with the poster child of professional drivers, and concluding with the privileged elites in the law and finance.

It's time to answer the question: will it be different this time? Will machine intelligence automate most human jobs within the next few decades, and leave a large minority of people – perhaps a majority – unable to gain paid employment?

Substitution and complementary forces

In his 2019 book, "A World Without Work", Daniel Susskind provides a helpful explanation of the two main economic forces that will determine whether and when there is technological unemployment: the substitution force and the complementary force.

The substitution force is straightforward: machines replace horses and humans in jobs if they are cheaper, better, and / or faster. In 1915 there were 21 million horses labouring away in America, and the US horse population today is two million. Machines substituted for horses.

Despite numerous rounds of automation (mostly mechanisation so far) humans have not been ejected from the workplace, and many developed economies are close to full employment. This is because the complementary force has outweighed the substitution force. Humans have been substituted out of many jobs, such as lift operator and secretary, but the complementary force has opened up new jobs for us. The complementary force works in three ways: the productivity effect, the bigger pie effect, and the changing pie effect.

The productivity effect is when automation eliminates some jobs, but makes other workers more productive. Imagine a shop with five cash tills manned by people. It replaces them with five self-checkout machines which are operated by customers, and one member of staff to supervise them. The one remaining staff member has become five times more productive.

The bigger pie effect is clearly visible in the economic history of the USA: its GDP was 15,000 times higher in the year 2000 than it was in the year 1700. This is partly because automation made the economy more productive, which means more wealth, more demand, and more jobs.

The changing pie effect is seen in the shift in economies and employment from farms to factories, and then to offices. Humans are substituted in one part of the economy, but new jobs are created in new parts.

So the big questions are: will the complementary force continue to work forever? And if not, when will it stop?

Jobpocalypse later

Prior to the Covid pandemic, the US and the UK were pretty close to full employment. True, there was an amount of joblessness disguised as self-employment, and labour participation rates had declined in some countries, especially in the US. It was also argued that many people were doing less rewarding jobs today than their counterparts a generation ago were doing. Nevertheless, recent and current employment data does not support a claim that permanent widespread technological unemployment is imminent.

If there is no data to suggest that technological unemployment is imminent, can we conclude anything about its likelihood in the future from the review of employment sectors in the previous three chapters? There are nay-sayers, but many people will agree that the number of professional drivers, warehouse staff, call centre operatives, retail staff and others will be significantly reduced a generation from now. But the complementary effect could find new jobs for those people.

The evidence is more mixed when we look at the professions. Machines will clearly take over more and more of the tasks currently performed by journalists, teachers, doctors and lawyers, but the iceberg effect could keep humans employed in those areas because the machines reduce the cost of writing, teaching, doctoring and lawyering, and thereby increase the demand for them.

It boils down to this. There will be plenty of jobs for humans so long as there are plenty of tasks which we can do for money, and machines cannot. But if and when the day dawns on which machines can do pretty much everything that we can do for money, then there will be lasting widespread technological unemployment.

That day will dawn if the cognitive abilities of machines continue to improve faster than ours do. And if the cognitive abilities of machines continue to improve at an exponential rate, that day may dawn sooner than most people think. We saw in chapter 13 that 20 years of exponential growth will produce an 8,000-fold improvement in the capabilities of machines, and 30 years will produce a million-fold improvement. Machines 8,000 times more capable than the ones we have today may well be able to do the vast majority of the tasks that we can do for money cheaper, better, and faster than us. Machines a million times more capable will almost certainly be able to do so.

So 20 to 30 years seems a reasonable time frame in which to expect technological unemployment – longer if the rate of improvement slows.

Optimism

People who think technological unemployment is probably coming are often described as pessimists. Some of them are: they are building secure hiding places in remote locations because they think social collapse is coming. But this is not the only possible response.

A world in which machines do most of the jobs could be a world in which humans get on with the important parts of life: playing, having fun, learning, exploring, being creative, playing sports, socialising, inventing things. We could have a second renaissance.

The real pessimists are those people who insist that all humans must continue doing jobs forever. Some people love their jobs all the time, but most do not. A 2017 poll by Gallup found that only 15% of employees worldwide are genuinely engaged in their work.[326] The rest do it to put food on the table. For most people, jobs give us a reason to get up in the morning, and they fill our time. They provide money, and perhaps status, but most of us could think of better things to do with our days. If we didn't have to worry about money, we could travel, read, paint, do sports, hang out with friends, play video games, learn a language. We could even help other people. Many of us – probably most of us – would work, at least some of the time. But we would not get paid for it.

Financial Times columnist Martin Wolf urged that we should "enslave the robots and free the poor",[327] and who would not welcome such an outcome? We will explore all this in more detail in chapter 23.

Education will have a very important role if we make it through successfully to the new world in which many or most of us are permanently and irrevocably unemployed. We will need good education to take advantage of our leisured lives even more than we did to survive and thrive in our working lives. But the education we will need will be vacational, not vocational.

The Churn

In the decades before technological unemployment arrives, AI-driven automation will have an increasingly disruptive effect on the job market. This will be the period of the great Churn. People will lose their jobs to machines, and they will have to be redeployed within companies, between companies, and between industries. With increasing frequency. Workers will have to learn how to work with machines: for a long time, AI won't replace humans in all jobs, but increasingly, humans who can work with AI will replace those who cannot. We cannot yet say what jobs people will be retrained to do; no doubt some of them will surprise us.

This process will be disturbing and frightening for many of us. Governments will need to support their people through the Churn with enhanced welfare programmes and other methods of economic support. The experience of massive government intervention during the corona virus crisis should teach us valuable lessons about what works and what doesn't.

During the period of Churn, we will need to re-train ourselves more and more frequently, and do it faster each time. Education and training are notoriously hard industries to reform. Here again, AI will provide solutions as well as raising the challenge in the first place. Education and training will become personalised, as we all acquire learning assistants - digital coaches which know better than we do (and better than any human teacher could do) what we already know, what we need to learn next, and how to optimise that learning process. For many years, these assistants will work alongside human teachers rather than replacing them. If we manage the transition smoothly, then by the time the replacement happens, we will hardly notice, and we won't object.

Phase change?

If and when technological unemployment does arrive, it may happen the way Hemingway described going bankrupt in his 1926 novel "The Sun Also Rises": "How did you go bankrupt?" Bill asked. "Two ways," Mike said. "Gradually and then suddenly."

Companies dislike sacking lots of people: it is bad PR, and bad for morale. Insofar as they can, firms will redeploy people whose jobs have been automated, and cut down on general recruiting. This will disguise the process for a while, but people will notice it is getting harder to find new jobs. Whenever there is a crisis, companies will take the opportunity to let some people go. Eventually they won't be able to hide what is happening, and the trickle will become a flow.

This is the phase change mode of technological unemployment, like the transition from water to steam through boiling. You have to inject a lot of heat into water before bubbles start to form, but once the phase change starts, its progress is rapid.

Alternatively, the process may be gradual and inexorable, with machines constantly chipping away at more and more areas of employment.

Confidence level

We cannot claim to know for certain that technological unemployment will happen. We don't have a crystal ball. But it seems likely, and it certainly seems sufficiently plausible that we should make plans for what to do if it does happen.

Some people are surprisingly vehement that it won't happen. For instance, in May 2017 Eric Schmidt proudly proclaimed himself a "job elimination denier," saying that the transformation we are going through will be no different to previous economic revolutions.[328] In the absence of a crystal ball, this attitude seems dangerous.

If technological unemployment never arrives, and we have spent a modest amount of resources contingency planning for it, then we will have lost very little.

On the other hand, it if does arrive and we have done nothing, the consequences could be grave. We should avoid that.

CHAPTER 22

SKEPTICAL ARGUMENTS

The Reverse Luddite Fallacy

As we saw in chapter 16, the Luddite Fallacy is the argument that we have had automation for centuries, but it has never caused lasting widespread unemployment, and therefore it never can. People have raised fears about unemployment many times down the years, and they have always proved false. They are like the boy who cried wolf in Aesop's fable.[329]

When the argument is expressed this baldly its weakness is clear: it is akin to saying that we have never sent a person to Mars so we never will. Or to someone in 1907 saying that despite numerous false alarms, we have not achieved powered flight in the past, and therefore we never will.

It is also historically inaccurate - if you count horses as employees. As we saw above, there were 21.5 million horses in America in 1915, mostly pulling vehicles;

now there are pretty much none.

As for crying wolf, it is worth remembering that at the end of the story, the wolf did turn up and kill all the sheep. Automation has been going on for centuries, and past claims that it was causing permanent widespread unemployment have been proven wrong. But we should not be complacent when there are good reasons to think that this time, it may be different.

Past rounds of automation replaced human and animal muscle power. That was fine for the humans who were displaced, as they could go on to do jobs which were often more interesting and less dangerous, using their cognitive faculties instead of their muscle power.

Intelligent machines are replicating and in many cases improving on our cognitive abilities. As we saw in chapter 17, what many of us do at work these days is to ingest information, process it, and pass it on to someone else. Intelligent machines are getting better at all of this than us. They are already better than us at image recognition; they are overtaking us in speech recognition, and they are catching up in natural language processing. Unlike us, they are improving fast: thanks to Moore's Law (which is evolving, not dying) they get twice as good every eighteen months or so.

To argue that technological unemployment cannot happen in the next few decades simply because it has not happened in the past is complacent and foolish: it is the Reverse Luddite Fallacy.[330] However, the skeptics do have some better arguments.

Inexhaustible demand

Some people argue that humans will not become unemployable because there is an inexhaustible well of potential demand. In an blog post in June 2014, Marc Andreessen wrote "This sort of thinking is textbook <u>Luddism</u>, relying on a "lump-of-labor" fallacy – the idea that there is a fixed amount of work to be done. The counter-argument to a finite supply of work comes from economist Milton Friedman — Human wants and needs are infinite, which means there is always more to do."[331]

There may be plenty of demand, but if a machine can always provide the supply cheaper, better and faster, it will be economically compelling to have the machine do the job instead of a human.

Icebergs

We saw in the last section that applying AI systems to the professions can open up the possibility of huge new classes of work. Projects that would previously have been uneconomic become feasible. More articles can be written, more situations can be scrutinised legally, more patients can be diagnosed. This has been called the paradox of automation, or the iceberg phenomenon: it was thought that junior professionals in particular were standing on thin ice, but it turned out instead that they are standing on top of a massive bulk of newly available work. Their positions begin to look secure again.

However, over time, the machines will continue to get faster and more effective, and work their way down the iceberg.

The human touch

Some observers think that our salvation from cognitive automation lies in our very humanity. Our social skills, and our ability to empathise and to care mean that we carry out tasks in a different way than machines. Machines are by definition impersonal, the argument goes, and this renders them unsuitable for some types of job.

David Deming, a research fellow at the US National Bureau of Economic Research, believes we are already seeing the implications of this. In a report published in 2015 he claimed that the fastest growth in US employment since as long ago as 1980 has been in jobs requiring good social skills. Jobs requiring strong analytical abilities but no social skills have been in decline – with the implication that they are already being automated.[332]

Unfortunately, it isn't true that humans want to deal with other humans whenever possible. The first automatic deposit machine, the Bankograph, was installed in a bank in New York in 1960, but it was rejected by its intended customers. Its inventor, Luther Simjian, explained that "The only people using the machines were prostitutes and gamblers, who didn't want to deal with tellers face-to-face." There were not enough of them to make the machines a worthwhile investment.[333] The first cash dispensing machine, or ATM, was installed in a bank in North London in June 1967. At first, people were hesitant to use it, but that changed when they realised they no longer had to queue for their cash, and they could access it when the banks were closed (which was most of the time, in those days). Very quickly, people showed a marked preference for the machine over the human bank teller.[334]

Nursing is an occupation long associated with caring people. Images of Florence Nightingale emoting as she nursed the wounded of the Crimean War are deeply ingrained in the profession's public image. But there is growing evidence that robots make perfectly acceptable companions for sick people, and are sometimes preferable to their human equivalents. The Paro is a robotic seal developed for use in hospitals. Cute-looking, with big black eyes and covered in soft fur, it contains two 32-bit processors, three microphones, 12 tactile sensors, and it is animated by a system of silent motors. It recharges by sucking on a fake baby pacifier.

The Paro cost $15m to develop; it distinguishes between individual humans, and repeats behaviours which appear to please them.[335] It has proved especially popular with patients suffering from dementia. As Shannon Vallor, a philosophy professor at Santa Clara University remarked, "People have demonstrated a remarkable ability to transfer their psychological expectations of other people's thoughts, emotions, and feelings to robots."[336]

Japan is the test-bed for the acceptability of robotic elder care. Thanks to a consistently low birth rate, it has a greying population, and it has long been resistant to mass immigration as a solution for labour shortages. Furthermore its population is generally technophile. It is probably no coincidence that Softbank, one of Japan's largest companies, now owns several of the world's leading robotics companies, including Aldebaran and Boston Dynamics. The evidence so far is that the acceptability is high. A manager at a Tokyo home for the elderly says, "A lot of people thought that elderly people would be scared or uncomfortable with robots, but they are actually very interested, and interact naturally with them. They really enjoy talking to them and their motivation goes up when they use the rehabilitation robots, helping them to walk again more quickly."[337]

So humans are happy to interact with machines in more situations than we might expect. Machines are also better than we might expect at "understanding" humans, and providing appropriate responses.

Robot therapist

The US Army has a big problem with post-traumatic stress disorder (PTSD) among veterans, not least because soldiers don't like to admit they have it. DARPA funds research at the University of Southern California to develop online therapy services, and an online virtual therapist called Ellie proved to be better than human therapists at diagnosing PTSD.[338]

There are two reasons for this. First, soldiers feel less embarrassed discussing their feelings with an entity they know will not judge them. In one test, 100 subjects were told that Ellie was controlled by a human, and another 100 were told that it was a robot. This second group displayed their feelings more openly, both verbally and in their expressions.[339]

Secondly, and perhaps more interesting, Ellie gleans most of its information about what is going on inside the soldier's head from his facial expressions rather than from what he says. When talking to a human therapist, the soldier may successfully "sell" the idea that there is nothing wrong. The human therapist listens closely to what he says, and may miss the subtle facial signals that contradict him. Counter-intuitively, people with depression smile just as frequently as happy people, but their smiles are shorter and more forced. Ellie is superb at catching this.[340]

Most people would probably agree with David Deming when he says that "Reading the minds of others and reacting is [a skill that] has evolved in humans over thousands of years. Human interaction in the workplace involves team production, with workers playing off of each other's strengths and adapting flexibly to changing circumstances. Such non-routine interaction is at the heart of the human advantage over machines." In a decade or two, we may have to start re-thinking that.

Made by hand

Another way that people have suggested that the human touch could preserve employment is that we will place a higher value on items manufactured by humans than on items manufactured by machines. It is hard to see much evidence of this in today's world outside some niche areas like hand-made cakes.[341] Not many people today buy handmade radios or handmade cars.

There are four reasons why people might prefer

products and services made by humans rather than machines: quality, loyalty, variation, and status.

If humans provide a better product or a better service than machines, then other humans will buy from them. But the argument of this book is that within a generation or so, machines will produce most goods and services cheaper, better and faster than we can.

Loyalty to our species might be a better defence. "Buy hand-made, save a human!" sounds like a plausible rallying-cry, or at least a marketing slogan. The past is not always a reliable guide to the exponential future, but it is a good place to start, and unfortunately it does not augur well for appeals to loyalty. In the late 1960s, Britain was feeling queasy as the Empire dissolved and Germany's economic power was returning. The "I'm backing Britain" campaign started in December 1967, trying to get British people to buy domestically manufactured products instead of imports. It fizzled out within a few months.[342]

Car manufacturing has long been symbolic of a nation's manufacturing virility. In the 1950s, Britain was the world's second-largest car manufacturer after the US, but in the 1960s its designs and build quality fell behind first its European rivals, and then the Japanese. Despite repeated appeals to buy British, sales declined and in 1975 the remaining national manufacturer, British Leyland, was nationalised. It never recovered, and Britain is now home to none of the major global car brands. (Fortunately, it has many innovative and thriving automotive design and component businesses working for foreign brands.)

Appealing to people to buy handmade items out of loyalty to one's species may not have a huge economic impact if machine-made items are better quality and much, much cheaper. And in a world of falling employment, most people are going to have to buy as efficiently as they can.

The third reason for buying from humans could be summarised by the phrase "artisanal variation". We like antiques because the patina of age gives them personality: each one is unique. The same goes for the original work of an artist, even if it isn't a Vermeer or a Rubens. But for most people, this is the preserve of luxury items, a few select pieces which we keep on display. Most of our possessions are mass-produced because they are much cheaper, disposable, and we can afford a better lifestyle that way.

We have seen this before, in the second half of the 19th century. With the industrial revolution in full swing, William Morris helped found the Arts and Crafts movement to produce hand-made furnishings and decorations. His concern was to raise quality rather than to reduce unemployment, but in practice he ended up making expensive pieces which only the rich could afford.[343]

Finally, some people may choose to buy goods and services from humans rather than machines for reasons of status. But by definition, this could only ever amount to a niche activity, and would not save most of us from unemployability.

Entrepreneurs

If machines take over the jobs that are repetitive, humans will look to do things that require creativity, intuition, and pursue counter-intuitive paths. One job title which fits that description is entrepreneur.

In my experience there are two types of entrepreneur. Both are resourceful, determined, and usually of above-average intelligence. The first and most common type is someone who works in an organisation which is doing something poorly. They notice this, and decide to offer a better version. They utilise essential skills and industry know-how acquired while working for the original organisation, and simply improve incrementally on what was being provided there. These people are talented and hard-working, but they also had the good fortune to be in the right place at the right time to spot the opportunity. If they had not been in that position they would have spent their careers working for other people, and because they are hard-working and bright they would probably have make a good fist of that.

The second type is destined to be an entrepreneur whatever circumstances life drops them into. They will never be happy working for someone else. They envision themselves in a future world which looks impossible to anyone else, but they choose to believe it and by dint of sheer force of will they make that future a reality. They will walk through brick walls to make it happen, and will probably be bankrupt more than once. They are charming, astonishingly energetic, and often rather hard to be around. In the words of LinkedIn founder Reid Hoffman, they are people who will happily throw themselves off a cliff and assemble an aeroplane on the way down.[344]

Both types of entrepreneur are rare, especially the second kind – which may be a good thing for the rest of us. In any case, this is probably not an occupation that is going to save large numbers of people from technological unemployment. The other thing to remember about entrepreneurship as a career is that while founding a successful startup is one of the most exciting things a person can do, most startups fail.

Centaurs

A computer first beat the best human at chess back in 1997. Deep Blue was one of the most powerful computers in the world when it beat Gary Kasparov; the match was close and the result was controversial. Today, a programme running on a smartphone could beat any human.

For a while, a very good human chess player teamed up with a powerful chess computer could beat a second chess computer playing on its own. Humans could undermine the game of a computer by throwing in some surprise moves which didn't make much sense in the short term, or by deploying an intuitive strategy. Matches between humans working with computers were called advanced chess, or centaur chess. Kasparov himself initiated the first high-level centaur chess competition in Leon, in Spain, in 1998.

Some people believe this phenomenon of humans

teaming up with computers to form centaurs is a metaphor for how we can avoid most jobs being automated by machine intelligence. The computer will take care of those aspects of the job (or task) which are routine, logical and dull, and the human will be freed up to deploy her intuition and creativity. Engineers didn't become redundant just because computers replaced slide rules. Kevin Kelly, founder of Wired magazine, puts it more lyrically: "machines are for answers; humans are for questions."[345] Pablo Picasso said something similar back in 1964: "Computers are useless. They can only give you answers."[346]

Unfortunately, the supremacy of human-machine centaur chess players did not last long. Today, a good chess-playing computer is not improved by being paired with a human.

The magic jobs drawer

If machines are going to take a great many, perhaps most, of our existing jobs, can we create a host of new ones – perhaps whole new industries – to replace them? Those who think we can point out that many of the jobs we do today did not exist a hundred years ago. Our grandparents would not have understood what we mean by website builder, social media marketer, user experience designer, chief brand evangelist, and so on. Surely, the argument goes, all these new technologies we have been talking about will throw up many new types of jobs that we cannot imagine today.

As the person probably most responsible for Google's self-driving cars, Sebastian Thrun is a man worth listening to on the subject. He is optimistic: "With the advent of new technologies, we've always created new jobs. I don't know what these jobs will be, but I'm confident we will find them."[347]

Unfortunately, past experience is (again) not as encouraging as you might think. My late friend Gerald Huff was a senior software engineer working in Silicon Valley, ground zero of the developments we are talking about. Nervous about the prospect of technological unemployment, he carried out a comparative analysis of US occupations in 1914 and 2014. Using data from the US Department of Labour,[348] he discovered that 80% of the 2014 occupations already existed in 1914. Furthermore, the numbers of people employed in the 20% of new occupations were modest, with only 10% of the working population engaged in them. The US economy is much bigger today than it was in 1914, and employs far more people, but the occupations are not new.

Of course, those of us who argue that it is different this time cannot rely on the historical precedent. It might be different this time, in that vast swathes of new jobs will be created – including jobs for averagely-skilled people, not just relatively high-skill jobs like social media marketing. But those who argue that we are falling for the Luddite fallacy cannot argue that history points to everybody getting new types of jobs which are more interesting and safer after a period of adjustment. It doesn't.

we were to create a host of new jobs, what might they be? Maybe some of us will become dream wranglers, guiding each other toward fluency in lucid dreaming. Others may become emotion coaches, helping each other to overcome depression, anxiety, and frustration. Maybe there will be jobs for which we have no words today, because the technology has not yet evolved to allow them to come into being.

It's not hard to imagine that virtual reality will create a lot of new jobs. If it is addictive as enthusiasts think it will be, many people will spend a great deal of their time – perhaps the majority of it – in VR worlds. In that case there will be huge demand for new and better imaginary or simulated worlds to inhabit, and that means jobs.

But does it mean jobs for humans? Although the credit list for the latest superhero blockbuster stretches all the way around the block as it names everyone involved in rotoscoping and compositing the hyper-realistic armies of aliens, the latest CGI technology also makes it possible for two teenagers with a mobile phone to make a film which gains theatrical distribution. Their increasingly powerful software and hardware allows Hollywood directors to conjure visual worlds of such compelling complexity that their predecessors would rub their eyes in disbelief, but it also allows huge quantities of immersive content to be developed by skeleton crews. There will probably always be an elite of directors who are highly paid to push the boundaries of what can be imagined and what can be created, but software will do more and more of the heavy lifting in VR production.

Not for the first time, the games industry shows

what is possible. A game called "No Man's Sky" was announced in 2014 which can provide far more imaginary worlds than you could visit in a lifetime. They are conjured up by algorithms and random number generators. Quite literally, you boldly go where no programmer or designer has gone before.[349]

The historical record should not make us confident that we will invent hordes of new jobs to replaces the ones taken over by machines. It might happen, but the firm belief that it will is little better than a blind belief that there is a "magic job drawer" that we can open when we need to, and thousands of exciting new jobs will fly out.

Artists

After all these apparently gloomy prognostications, here is a more optimistic observation. There is one profession which probably cannot be automated until the arrival of an artificial general intelligence which is also fully conscious. That profession is art, and to understand why, we need to recall the distinction we made in chapter 6 between art and creativity.

While reviewing the progress that AI has made to date, we argued that machines can be creative. In mid-2015, Google researchers installed a feedback loop in an image recognition neural network, and the result was a series of fabulously hallucinogenic images.[350] To deny that they were creative is to distort the meaning of the word.

We went on to define art as the application of creativity to express something of personal importance to the artist. It might be beauty, an emotion, or a profound insight into what it means to be human. (We accepted without regret that this might disqualify a good deal of what is currently sold under the banner of art.) To say something about your own experience clearly requires you to have had some experience, and that requires consciousness. Therefore, until a conscious artificial general intelligence (AGI) arrives, AIs can be creative but not artistic.

This means that Donna Tartt and Kazuo Ishiguro are probably OK for quite a few decades. However, today's successful genre writers who use stables of assistants to churn out several crime and romance novels each year have chosen the timing of their careers expertly, and within a decade or two, their publishers may need to find something different to do.

Education

It is surprising how many smart people think that education is the "silver bullet" answer to automation by machine intelligence. Microsoft CEO Satya Nadella said in January 2016: "I feel the right emphasis is on skills, rather than worrying too much about the jobs [which] will be lost. We will have to spend the money to educate our people, not just children but also people mid-career so they can find new jobs."[351]

Massive Open Online Courses, or MOOCs, are promoted as the way we will all re-train for a new job each time a machine takes our old one. MOOCs are important, and along with flipped lessons, competency-based learning, and the use of Big Data, they will improve the quality of education, and make excellent learning opportunities available to all.

With flipped lessons, students watch a video of a lecture for homework, and then put what they have been told into practice in the classroom. The teacher acts as coach and mentor, a more interactive role than lecturing. Competency-based learning requires students to have mastered a skill or a lesson before they move on to the next one; students within a class may progress at different speeds. Big data enables students and teachers to understand how well the learning process is going, and where extra support is needed.

These techniques are exciting and powerful, and they will be invaluable during the Churn phase which we will hear about in the next chapter. But in the long run, they won't protect us from technological unemployment. We have seen that machines are increasingly capable of performing many of the tasks currently carried out by highly educated, highly paid people. In the longer term, the machines aren't just coming for the jobs of bricklayers; they're coming for the jobs of surgeons and lawyers too.

Timing

When pushed, many skeptics will agree that machines will take over pretty much all our jobs - just not in our lifetimes, nor those of our children and grandchildren. The skeptics might change their minds if they took exponential growth more seriously. As we saw, thanks to this progress, the machines we have in ten years' time will be 120 times more powerful than the ones we have today. In 20 years time they will be 8,000 times more powerful, and in 30 years, a million times more powerful. Those machines will surely take most of our existing jobs.

As well as not taking exponential growth into account, skeptics are often simply not thinking far enough ahead. It is easy to be complacent about machines not replacing humans in jobs if you only think five or ten years ahead. In that timeframe they won't be capable of replacing us in most jobs. For this reason I propose a new rule - Calum's Rule: every forecast about technology should specify the time horizon.

CHAPTER 23

CHALLENGES: MEANING AND INCOME

Challenges

The point of this book so far has been to persuade you that within a few decades, it is likely that many people will be rendered unemployable by machine intelligence. If I have not wholly succeeded in that aim, then I hope you are at least prepared to accept that the possibility is serious enough that we should be thinking about the implications, and what to do about it if it happens.

If I haven't even got you that far, then you're probably about to put this book down. If so, don't throw it away – you might want to come back to it when self-driving vehicles start to make serious impacts on the employment data.

If I have made the case successfully – or if you were persuaded before we started – then welcome to the next stage of the journey. If we are en route to a leisure society, what do we have to do to make the destination wonderful and the transition smooth?

There will be challenges. I anticipate six: meaning, economic contraction, income, allocation, cohesion, and panic. Let's take a look at each of them in turn.

Meaning

The meaning of life...

... is 42, of course.[352]

OK, now we've got that out of the way, would you agree with the statement that people's lives need to have meaning in order for them to feel fulfilled, satisfied, and happy? It's certainly true for me, and I'm pretty sure it's true for most of the people I know. It is probably also true of you, or you wouldn't be reading this book. The initial reaction of many people when they first take seriously the possibility of widespread lasting technological unemployment is, "how will we fill our days? How will we find meaning in our lives without work?"

I have met people who claimed to be pure hedonists – interested only in immediate pleasure. Some of them may even have been telling the truth. But most of us get bored if we feel our lives have no meaning. And not just bored in the sense that you get bored in a queue at a supermarket checkout, but profoundly restless and frustrated. To avoid this feeling we make deep emotional investments in ideas and institutions like family, friendships, work, loyalty to tribes (including political movements and football teams), nations and causes. Deprived of these things, we feel lost and alienated.

Perhaps the most famous quote attributed to the 4th century BC Greek philosopher Socrates is that the unexamined life is not worth living. It is a remarkably strong statement. Why not just say that an unexamined life – a life without philosophy, in other words – is less good than an examined one? Is an unexamined life really worse than death? He made the statement at his trial, when the outcome was a choice between exile and suicide (he chose suicide), so perhaps he was under stress and being hyperbolic. But the claim is usually taken at face value, and perhaps he meant it literally.

It is also an elitist statement. Many people are too preoccupied with making a living, raising a family, escaping drug addiction or whatever immediate challenge they face to indulge in the luxury of philosophical discourse. Are their lives not worth living? You could argue that Socrates and his fellow ancient Athenians had slaves to take care of the menial stuff, but we have labour-saving devices instead, so that's no excuse.

Of course the question of what constitutes a good life, a worthwhile life, a life with meaning is a vexed one, with no simple answers, and probably no single answer. The philosopher John Danaher distinguishes between subjective accounts, which involve feeling worthwhile, and objective accounts, which involve helping to make or do something worthwhile.[353]

Despite not knowing (or at least not agreeing) what

a meaningful life is, and despite not spending all that much time in the average day thinking about it, most of us believe we need it. And many of us find it in work. So it's going to be a problem if we stop working.

Or is it?

Meaning and work

Simon Sinek has made a name for himself with books that propound a simple but important truth: if you have a clear purpose which inspires others, you can achieve great things. His best-known saying is "Working hard for something we don't care about is called stress; working hard for something we love is called passion."

You could be forgiven for thinking that a law was passed a few years ago in the US requiring business leaders – and people who want to be business leaders – to talk about their passion for their business. But most people don't feel passionate about their work, even if they pretend they do. In fact, many people are positively alienated by their jobs. They find them meaningless and boring.

Yet even these people usually define themselves by what they do for a living. If you ask someone at a party what they do they are likely to reply that they are an accountant, a taxi driver or an electrician. They are less likely to say that they are the coach of their child's football club, or a cinema-goer, or a reader. No doubt this is partly due to the amount of time that our jobs absorb – but then again we don't define ourselves as sleepers. It also has to do with work being the activity that provides our income, which is why home-based parents often feel sheepish about naming that as their work. (They shouldn't: parenting is one of the hardest but most rewarding jobs you can do!)

So jobs help to define us, and they give many of us purpose. They even give some of us meaning. So how damaging would it be if we lost them? Unemployed people often struggle with depression, but they are experiencing it in the context of a society where it seems that everyone else has a job. They are also usually on a lower income than the employed people around them. How bad would unemployment be if everyone was unemployed, and receiving a decent income?

Fortunately, there are a couple of places we can look for an answer to that question.

The rich and the retired

The agricultural revolution, which happened in different parts of the world around 12,000 years ago, created sustainable surpluses of food and other basic resources. This enabled a class of people to stop doing the work that pretty much all humans had done since our appearance on the planet, which was foraging and hunting for food. They became tribal leaders, kings, warriors, priests, traders and so on. Sometimes they spent as much time on these activities as the people who continued to forage and hunt, but sometimes they took time off – deliberately or by happenstance – and engaged in lives of leisure.

Eventually, in Europe, these people became known as aristocrats, from a Greek word meaning the best –

originally in a military sense, and then in a political one. Some aristocrats did jobs: they ran their estates, they got involved in politics, and in some countries they ran empires. Occasionally they became men (and more rarely, women) of what we now call science. Famously, they disdained trade and commerce, regarding those activities as the preserve of the class below them, the middle class.

Many aristocrats did not work – including almost all the female ones. They led lives of leisure. Starting in the late 17th century, young aristocratic men (and in a few cases, young aristocratic women) toured classical Greek and Roman sites in the Mediterranean countries. Returning home, they mostly socialised. Their lives revolved around balls, hunts, and visits to their local peers, interspersed with the glamour and tragedy of war, if that was their inclination. This lifestyle was chronicled in the novel, an art form which first acquired its current realistic form in the early 18th century.[354]

The lives depicted by Jane Austen and her contemporaries may seem tame to modern readers, who have experienced international travel and expect simultaneous global communications. But they were agreeable lives compared to what their poorer contemporaries had to put up with. Addictions to gambling and drink were a hazard, and of course a minority of this pampered class destroyed themselves and their families with these vices. But this was unusual, and by and large most 18th- and 19th-century European aristocrats seem to have passed their lives without great concern about their lack of meaning. Whether these lives were worthwhile or not, whether or not they had meaning, is probably not for us to judge, but there is no evidence of widespread existential angst among the nobility.

In fact, it is these privileged people who made most of the advances in human thought and art in previous centuries, precisely because they did not need to work for a living, or eke out an existence as subsistence farmers. If they did not produce the memorable work themselves, they often sponsored it by employing talented artisans. It seems there is something to be said for the ability to be idle.[355]

The other group we can look to for evidence about the effects of joblessness are retired people living on good incomes today. The conventional wisdom used to be that growing old was an almost unmitigated disaster: "Old age ain't no place for sissies", as Bette Davies said,[356] although it's obviously better than the only alternative currently available. But starting in the 1990s, researchers began questioning this perception, and found instead that the progress of happiness throughout life is U-shaped. We are at our happiest and most fulfilled when young, we become stressed and discontented in our prime and middle age, and we are happier and more relaxed again when older, despite the onset of physical disabilities and limitations.[357] This pattern has been observed across a wide range of societies, and over a substantial period of time.

There are probably numerous causes of this effect, including escape from the responsibility for looking after children, and the acquisition of wisdom, including an acceptance of what life has thrown at us. But the absence of jobs plays a major role in the lives of the retired. Even if it is not causing the up-tick in happiness, it is at least not preventing it.

Retiring in penury is no fun at all. But if you spend a few days in the towns and villages where better-off people retire – perhaps on defined benefit, final salary pensions – you will see very busy communities of people organising festivals and dinner parties, volunteering for worthy causes, shuttling between games of bridge and lessons at the university of the third age. If you ask them how they manage to fill their days, they will tell you they have no idea how they used to find time for work.

Virtually happy

Thus far in human history we have had to find our meaning within the constraints of the three-dimensional world we live in, or in our imaginations. Technology is poised to open up a whole new space for us to explore together – the world of virtual reality. We don't yet know how we will react to this new universe, how we will behave in it, and what it will mean to us. We can be pretty confident that it will have a big impact.

"Diaspora", Greg Egan's novel of the far future, features an environment called the Truth Mines. It is a physical representation of mathematical theorems (albeit in virtual reality) which can seemingly be explored forever without exhausting all the discoveries that can be made. The ability to create virtual worlds that are so convincing to our brains that we almost lose the understanding that they are artificial may well allow us to expand enormously the space within which we find happiness and meaning.

Helping us adjust

As we saw above, only one in eight people around the world report being positively engaged in their work. But for the many people whose lives are given purpose – if not meaning – by their jobs, there may be some help required in adjusting to a jobless future. Many will need a little help to start, and then they will be fine. Others will continue to need periodic doses of support. Some will stumble badly, and will require extensive and extended assistance. We may have to do a lot of research and experimentation to get this right. But the examples of the retired and the aristocrats suggest that loss of meaning will not be one of the biggest problems that widespread technological unemployment will create.

Economic contraction

American union boss Walter Reuther recounts a story about a visit he made in the 1950s to a Ford manufacturing plant, where he saw an impressive array of robots assembling cars. The Ford executive who was showing him round asked how Reuther thought he would get the robots to pay union membership fees. Reuther replied that the bigger question was how the robots would buy cars. (The story is usually told with Henry Ford II playing the role of the company executive, but it almost certainly wasn't.[358])

The basic economic problem which this story is supposed to illustrate is that if nobody is earning any money then nobody can buy anything, and even those who do have money and resources can't sell anything. The economy grinds to a halt and everybody starves.

Of course life is never as black-and-white as that. Economies rarely go overnight from functioning tolerably well to complete collapse. Even catastrophic decline is less like falling off a cliff and more like tumbling down a slope, with pauses along the way as you hit ledges. But obviously, severe economic contraction is grim, and to be avoided if at all possible.

If and when machine intelligence renders more and more people unemployable, then other things being equal, the purchasing power previously exercised by those people will dry up. Their productive output will not be lost – it will just be provided by machines instead of humans. As demand falls but supply remains stable, prices will fall. At first, the falling prices may not be too much of a problem for firms and their owners, as the machines will be more efficient than the humans they replaced, and increasingly so, as they continue to improve at an exponential rate. But as more and more people become unemployed, the consequent fall in demand will overtake the price reductions enabled by greater efficiency. Economic contraction is pretty much inevitable, and it will get so serious that something will have to be done.

But before policy makers are forced to take action to tackle economic contraction, they will be faced by a much more serious problem: what to do about all those people who no longer have a source of income? Tackling this successfully will also solve the problem of economic contraction, so we can move right along.

Income

At the height of the Great Depression in the early 1930s, unemployment reached 25% of the working-age population.[359] Social security arrangements were primitive then, and developed societies were much poorer than they are today, so that level of joblessness was much harder on people than it is today, when parts of Europe have returned to similar levels overall,[360] with youth unemployment hitting 50% in some places.[361]

The worst levels of unemployment in developed countries today are found in Mediterranean countries like Greece and Spain, where family networks remain strong enough that sons and daughters can be supported for months or even years by fathers and mothers – and vice versa. There are escape valves, too, for the social pressure created by the situation. Economies further north are struggling less, and can absorb the energies and ambitions of many of the unemployed young people from the south.

When self-driving vehicles and other forms of automation render people of all classes unemployed right across the developed world, these safety nets will no longer be available. Articulate, well-connected and forceful middle class professionals will be standing alongside professional drivers and factory workers, demanding that the state do something to protect them and their families.

CHAPTER 24

UNIVERSAL BASIC INCOME

UBI

If and when societies reach the point where we have to admit that a significant proportion of the population will never work again – through no fault of their own – a mechanism will have to be found to keep those people alive. An answer which has become increasingly popular in the last few years is a Universal Basic Income (UBI), available to all without condition; a living wage which is paid to all citizens simply because they are citizens.

Probably the longest-standing organisation advocating UBI is the Basic Income Earth Network. BIEN was formed as long ago as 1986, and "Earth" replaced "European" in its name in 2004. BIEN defines UBI as "an income unconditionally granted to all on an individual basis, without means test or work requirement." UBI has also been called unconditional basic income, basic income, basic income guarantee (BIG), guaranteed annual income, and citizen's income.

Proponents have argued for various levels of UBI, but in general they choose a level at or around the poverty level in the country of operation. This is partly because they understand that a higher level would be (even more) unaffordable, and partly to ward off criticisms that UBI would make people lazy and unproductive.

The benefits claimed for UBI address issues which concern both the political left and right. Left-wing proponents see it as a mechanism to eradicate poverty and redress what they view as growing inequality within societies. They sometimes argue that it tackles the gender pay gap, and redistributes income away from capital and towards labour. It has also been held out as a partial solution to the alleged generational theft whereby relatively wealthy pensioners are receiving income generated by taxes on young workers who have no assets, and who may not themselves receive similar benefits in later life because the welfare system looks increasingly unaffordable.[362]

Right-wing advocates see UBI as a way to remove swathes of government bureaucracy: abolishing means testing removes the need for the battalions of civil servants who devise and implement it. There would be no incentive for people to game the benefits system, thus reducing government-generated waste and unfairness. They hope it would facilitate a wholesale simplification of tax structures, and perhaps enable a move to a flat tax. And they argue that more lower-income people would go to work because they would no longer be caught in benefit traps which penalise them for raising their income slightly. This would mean fewer children raised in families where nobody works, a particular bugbear of the right.[363]

Most current supporters of UBI are on the left, but it has had support from prominent right-wing politicians and economists in the past, notably President Richard Nixon, and economists Friedrich Hayek and Milton Friedman.

Experiments

There have been a surprising number of experiments with UBI: the Basic Income page on Reddit lists 25,[364] and gives potted descriptions of the purpose and outcomes of six of them.[365] All the researchers involved reported excellent results, with the subjects experiencing healthier, happier lives, and not collapsing into lazy lifestyles or squandering the money on alcohol or other drugs. Given that, it is curious that none of the experiments have been extended or made permanent.

The declared purpose of many UBI experiments is to investigate the concern that when people receive money for nothing, they stop working. One of the biggest experiments conducted so far, involving all 10,000 people in the small town of Dauphin in Manitoba, Canada, found that the only two social groups which did stop working were teenagers and young mothers, and this was seen as a positive outcome.[366]

Of course, people handing in their notice will be of no concern if machines have already stolen all our jobs, but a more subtle version of the concern remains: do people in receipt of money for nothing stop doing anything of value? Do they become indolent couch potatoes, watching TV all day long, or collapse into reliance on alcohol and other drugs? Bearing in mind the distinction we made earlier between jobs and work, in a world where intelligent machines have automated most economic activity, the question is not, do people give up jobs, but do they give up work?

Unfortunately, none of the UBI experiments carried out so far constitute a rigorous test. A rigorous test would be universal, randomised, long-term, and basic – in the sense that the income distributed should be enough to live on.[367] And so more tests are planned.

In fact, a number of significant UBI experiments are planned or under way at the time of writing. One in particular, in Finland, caused great excitement when it was announced in 2015, but the aims related to the right-wing concerns listed above. The Finnish researcher in charge of designing it, Olli Kangas, was hoping to demonstrate solutions for three problems with the current Finnish benefit system. First, people working part-time (perhaps in the gig economy) receive neither work-based benefits nor unemployment benefits. Second, some people are caught in a benefits trap whereby as their income increases their benefits decrease, which removes their incentive to work more and contribute more to the economy. Third, the existing benefits system is expensive, requiring too many bureaucrats to administer it.

The sample of Finns who were chosen to receive

the UBI were compared with a control sample who were not. Kangas explored their propensity to continue working, their reported happiness and well-being, and any changes in their use of health and social services. He had hoped to recruit a substantial sample – perhaps 100,000, which would enable him to detect variations between people of different ages, locations, demographics, and employment histories.[368] In the event, the trial was scaled back to just 2,000 people, who each received a modest €560 per month.

The results were reported in May 2020. The payments did not do much to encourage recipients into work but they did improve their mental well-being, confidence and life satisfaction.[369]

There is no shortage of places keen to experiment with UBI. The Dutch cities of Utrecht, Groningen, Wageningen and Tilburg are asking their national government for permission to carry out trials. All these initiatives are looking for ways to tackle problems with existing social welfare systems.

We have to go to Silicon Valley to find an experiment specifically designed to explore the impact of UBI in the context of a jobless future when machine intelligence has automated most of what we currently do for a living. Just such an experiment was announced in January 2016 by Sam Altman, president of the seed capital firm Y Combinator, which gave a start in life to Reddit, AirBnB and DropBox. The plan was to select 3,000 individuals at random from two US states. 1,000 of them will receive $1,000 per month for five years, and for comparison purposes, the other 2,000 will receive $50 a month.[370] At a total cost of $66m, this is a serious project, and it suffered from serious delays. In August 2018 it was announced that the project might start the following year,[371] but in June 2020 it was still in preparation by a new research organisation spun out from Y Combinator.[372]

Socialism?

With all these experiments bubbling up, the concept of UBI has become a favourite media topic, but it is controversial. Many opponents – especially in the US – see it as a form of socialism, and the US has traditionally harboured a visceral dislike of socialism. Martin Ford's book "The Rise of the Robots" expressed the hope that UBI was part of the solution to technological unemployment, but his hope was tarnished by his fear that "guaranteed income is likely to be disparaged as 'socialism'", and introducing it will be a "staggering challenge".

The strong performance of Senator Bernie Sanders, a self-proclaimed democratic socialist, in the Democratic Party's primaries in 2016 and 2020 show that there is a market for socialism in the USA today, but still a minority one. Of course, public opinion does shift, sometimes in response to a crisis. If and when it becomes impossible to deny that a majority (or even a large minority) of its citizens will never do paid work again, and for no fault of their own, it is unlikely that the rest will simply allow them to starve.

The dramatic recent changes in American attitudes towards homosexuality and drugs show how fast opinions can change, and how far. As recently as 1962, homosexual acts were illegal in every US state, and it was only in 2003 that the federal Supreme Court decision in the Lawrence v Texas case invalidated the ban in the last 14 states where it remained unlawful. (Even today, more than a dozen states have yet to repeal or amend their own legislation to reflect this ruling.[373]) And yet in June 2015, the federal Supreme Court ruled that bans on same-sex marriage are unconstitutional, in the case of Obergefell v Hodges. According to a Wall Street Journal poll, public support for gay marriage has doubled in the last decade, standing now at 60%.[374]

Attitudes towards the legalisation of cannabis have also undergone a rapid sea change. For years, governments proclaimed a war on drugs, but that policy has clearly failed. Billions of dollars have been spent, and countless lives have been lost, but supply has not been constrained, much less eliminated. Parts of Mexico and other countries where the drugs are grown or routed have become war zones, and hugely powerful criminal organisations have been spawned. Attempts to curb demand have also failed, with tens of thousands of people being criminalised for an activity that harmed no-one. Drugs are dangerous, and their supply should be regulated, but ceding control over that supply to criminal gangs has not proved an enlightened policy. Public opinion in America is swinging rapidly towards that position. In 1969 only 16% of voters polled by Gallup supported legalisation, but now a majority takes that view.[375] At the start of 2020, possession of cannabis for personal recreational use was legal in 11 states,[376] with the federal government agreeing not to interfere.[377]

It is not only America which is experiencing revolutionary changes in social attitudes. Up until 1997, sex before marriage was illegal in China, and condemned as "hooliganism". Nevertheless, a researcher found in 1989 that 15% of citizens had experienced it. The percentage had risen above 70% by 2014. Homosexuality was illegal until 2001, and gay marriage is still not legal. But in 2011, state-owned media began writing positive articles about gay pride marches in Shanghai and elsewhere.[378]

These examples show that entrenched societal opinions can and do change, sometimes quickly. If and when machine intelligence renders many of us permanently unemployable, it seems reasonable to expect that opposition to some form of payment to the long-term unemployed will evaporate. After all, the alternative would be mass starvation, and almost certainly, social collapse.

Inflationary?

Opponents of UBI also worry that it will stoke inflation. Other things being equal, a massive injection of money into an economy is liable to raise prices, leading to sudden inflation and perhaps even hyper-inflation. But as UBI campaigner Scott Santens points out, UBI does not necessarily mean an injection of fresh cash into the economy. Proponents believe it could be paid for by increased taxation of the better-off, and by replacing the existing benefits system, together with the bureaucracy which implements it.[379] Santens claims that where basic incomes have been introduced, as in Alaska in 1982 and Kuwait in 2011, inflation actually fell.

Unaffordable?

An objection to UBI with more substance is that it is unaffordable. Some argue that it can be funded by raising taxes on the small minority who have become extraordinarily wealthy in recent years. After all, even some of those wealthy people themselves (like Bill Gates and Warren Buffet) have confessed to feeling under-taxed.

Very wealthy people, especially in America, often decide to dedicate much of their wealth to charitable causes. Bill Gates (again) and Mark Zuckerberg are obvious examples, and many of the robber barons of the late 19th century gave fabulous sums to charitable foundations.

(The super-rich might decide to surrender much of their wealth in the interest of self-preservation as much as for philanthropic reasons. We might even call the tax the NAWCR tax, the Not being first Against the Wall Come the Revolution tax[380]. Alternatively they might be skeptical about how grateful the rest of us are likely to be, and decide instead to live on floating fortresses, protected by AI-powered defensive systems of awesome capability.)

If people do give their money away, they generally want to determine for themselves how their wealth is deployed, not least because they believe that they will make better use of it than politicians and bureaucrats.

So even the most generously disposed wealthy people often resist the wholesale appropriation of their assets in the form of taxation. And as demonstrated by the Panama papers scandal that erupted in April 2016, they are well equipped to do so, either by hiring clever lawyers and accountants to find loopholes and dodges, or by shifting themselves and their assets to less demanding jurisdictions.

Furthermore, entrepreneurs and other capable commercial people who are not yet extremely wealthy but aspire to become so, may decide to move out of a jurisdiction which raises taxes sharply to pay for UBI. Or if they stay, they may become discouraged and decide against taking the necessary risks and dedicating the necessary time and energy to projects which could achieve their ambition. These people are responsible for much of the dynamism in capitalist countries, and dampening their enthusiasm or incentivising them to move elsewhere can be very damaging to an economy.

This sounds like common sense, but is in fact highly contentious. The political left believes that inequality is a social evil, and argues that taxing the rich does not deter economic activity.[381] The political right believes that a modicum of inequality is no bad thing, and is anyway inevitable in a thriving economy. It argues that increasing taxes on the rich does deter economic activity, and may actually result in lower government revenues, as the rich look harder for ways to reduce their tax burden.[382]

The Laffer Curve

Unfortunately, the data is muddy, which enables both sides to marshal apparently convincing arguments. And as is so often the case, the truth lies somewhere between them. We do know that there is a level of taxation beyond which further increases are ineffective, or even self-defeating. At 100%, no-one would work, so that is an inefficient rate; 99% would not be much better. The Laffer Curve plots tax rates against the revenue they raise, but sadly we just don't know for sure what the optimal level is, either in general, or in a specific country at a specific time.[383]

In the UK, the Labour government in 2010 introduced a top rate of tax of 50% for people earning above £150,000. The Conservative government took it down to 45% in 2013, and claimed the result was a sharp revenue increase. The Labour party, of course, claimed the opposite.[384]

Which side you choose in this debate will be determined largely by your political orientation. People on the right believe that competing organisations in well-regulated markets are more efficient and effective than monopolistic governments, and that lower tax regimes encourage entrepreneurship. People in the centre agree with this, but also think that government spending is necessary to maintain a strong economic safety net for the less commercially capable and the less fortunate. These groups also think that governments tend to tax their subjects as much as they think they can get away with, which explains why so much of their tax take is achieved through subtle, indirect, and often downright stealthy taxes. They think that a substantial tax increase to fund UBI is likely to be economically damaging.

If you are on the political left you are likely to disagree with much of that. But proponents of UBI think they have two potential sources of revenue to pay for the scheme without "soaking the rich" (an expression coined in 1935 when Franklin D Roosevelt raised the top rate of income tax to 75% in order to pay for the New Deal.) These are saving money by eliminating bureaucracy, and taxing robots.

Let's kill all the bureaucrats

Channelling Shakespeare,[385] UBI advocates claim that the massive cost of UBI could be offset by abolishing much or all of the existing benefits systems, along with the legions of bureaucrats who implement them. They offer an enticing vision of a world without means-testing, with no poverty traps, no steely-eyed "advisers" in job centres forcing claimants to apply for unsuitable work, no benefit fraud and no need to game the system.

Unfortunately the world probably won't allow such a nice, tidy outcome. People's needs vary according to their capabilities, their life stage, and their location, among other things. Someone who is disabled might well suffer greatly if their income was equal to that of an able-bodied person in robust health. A single dad with a child may need extra support. People living in London or San Francisco would certainly need more housing benefit than people living in Albuquerque or Auchtermuchty. Having ushered all the bureaucrats out the door thanks to the purifying simplicity of UBI, we would have to apologise and invite them right back in again.

The RSA, a British think tank, published a report about UBI in December 2015 which was the result of a year's research and discussions.[386] It proposed abolishing much of the UK's existing benefits system, and replacing it with a payment of £3,692 for everyone between 25 and 65. This is £307 a month, £71 a week, or £10 a day. The payment amounts to a very modest 14% of the average UK wage, which was £26,500 in 2015.[387] People aged between 5 and 25 would receive £2,925, and pensioners would get £7,420. Extra payments would be made for young children.

The RSA estimated the total cost of its proposed system at £280bn, including annual running costs of £3bn. It claimed that this would be offset by £272bn saved by abolishing most of the existing benefits and pensions infrastructure, including personal income tax allowances and tax relief on pensions payments for higher rate tax payers.

The RSA claimed that families with children and on low wages would be £2,000 to £8,000 better off per year because of the removal of benefit traps. Adjustments required to prevent poorer people being worse off would take the cost to between £10-16bn, around 1% of GDP. This would be funded by taxes on high earners, a group which would also lose income from the changes.

The RSA scheme was not a fully-fledged UBI proposal, as payments would taper off for incomes above £75,000, and stop altogether at £100,000. The payments were also set at a level which would keep people alive, but would not provide a decent standard of living. It was also significant that the proposal ignores payments for housing and disability, which are of course substantial, and would require the recall of at least some of those bureaucrats.

Like the Finnish experiment, the idea makes more sense as an attempt to simplify and streamline the UK's messy and byzantine benefits system.

Countries are not isolated economic ecosystems. Introducing UBI would significantly affect the competitive position of a country which introduced it, and would have other unintended consequences. A surprisingly positive article about UBI in the right-of-centre Daily Telegraph newspaper in December 2015 speculated that if Finland's UBI experiment was successful, it would be inundated by economic migrants unless it leaves the EU.[388] Sam Altman has observed that implementing UBI in the USA would require very strict limits on immigration.[389]

Taxing robots

If we can't fund UBI by soaking the rich, or by sacking the bureaucrats who operate the existing welfare infrastructure, can we do it by taxing robots?

In an article published in February 2017, Bill Gates floated the idea of taxing robots which replace human workers.[390] The article was accompanied by a short video, at the end of which Gates giggled about the idea of paying more taxes. It got a lot of people very excited, but it is pretty certain that it won't work, and Gates probably understands this. (After all, he is a rather smart fellow.)

Imagine two firms offering the same service. One has been going for a few years, and recently replaced a thousand humans with machines. The other is a start-up, and went straight to machines. The former would be hit by a tax that the latter would escape. Not only is that very unfair, it would simply mean the former would close, and the tax take would disappear anyway.

Machines will rarely replace humans on a one-for-one basis. Humans will disappear from call centres, for instance, and be replaced by an AI system running on large numbers of servers in a big building with great air conditioning. Does the government tax one entity – namely the AI system - or each individual server? Or does it estimate the number of humans who have been laid off by the system and calculate the tax based on that?

In reality, Gates was probably offering a metaphor for how to cope with the economic singularity. If humans are being laid off in their millions by AI systems, somebody somewhere is making a lot of money by providing the technology which makes that possible. If – and it's a very big if – we can stop them skipping off to the Cayman Islands or somewhere else with very low tax rates, then perhaps we can levy taxes on them to pay an income to the people who have been laid off.

Unfortunately, even this more generalised approach

has fatal flaws. The technology replaced the humans because it was cheaper. (It may also produce better products and services and do so faster, but lower cost is almost always going to be part of the reason why these systems are adopted.) That means there is less profit being generated overall, which means a smaller tax base. And as the machines get more and more efficient, that tax base erodes faster.

The erosion of the tax base doesn't stop there. If a company makes a big profit by replacing a group of workers (albeit less money is now circulating through that part of the economy because it is has become more efficient) then competitors will attempt to replace it by making better AI systems. Assuming they succeed – which they often will – they will depress the tax base further.

Perhaps Gates saw this clearly too, and perhaps his real purpose was simply to get more people thinking and talking about how to provide the income that people will need after we lose our jobs.

Too low, or too expensive

It's looking hard to fund UBI by soaking the rich, eliminating the bureaucrats, or taxing the robots. FT journalist Tim Harford wrote that in current circumstances, UBI appeals to three kinds of people: those happy to see the needy receive less income, those happy to see the state balloon (and risk massive capital flight), and those who can't add up.[391]

John Kay, economics professor at the London School of Economics, wrote in March 2017 that "either the basic income is impossibly low, or the expenditure on it is impossibly high."[392]

Two out of three ain't good

As well as being unaffordable, the concept has another couple of serious problems which feature in its name: universality, and being basic.

In 1977 the American singer Meat Loaf released the classic album "Bat out of Hell", and the phrase "ear worm" could have been invented for one of the singles from it, "Two out of Three Ain't Bad". UBI has three components: it is universal, basic, and an income. As Meat Loaf might say, two out of these three ain't good.

Universal

The first of UBI's three characteristics is its universality. It is paid to all citizens regardless of their economic circumstances. There are several reasons why proponents want this. Experience shows that many benefits are only taken up by those they are intended for if everyone receives them. Means-tested benefits can have low uptake among their target recipients because they are too complicated to claim, or the beneficiaries feel uncomfortable about claiming them, or simply never find out about them. Child benefits in the UK are one well-known example. There is also the concern that UBI should not be stigmatised as a sign of failure in any sense.

But in the case of UBI, these considerations must be offset against the massive inefficiency of universality. In a scenario of massive unemployment, paying UBI to

Rupert Murdoch, Bill Gates, and the millions of others who are still earning healthy incomes would be a terrible waste of resources.

Basic

The second characteristic of UBI is that it is Basic, and this is its biggest problem of all – even worse than its unaffordability. "Basic" cannot mean anything other than extremely modest, and if we are to have a society in which a very large minority or a majority of people will be unemployable for the remainder of their lives, we have to do better than providing them with a subsistence income.

This isn't just a question of social justice, important though that is. It is also a question of self-preservation for the people who are still earning. A society in the developed world where a large minority of people have gone from a reasonable standard of living to a subsistence income which they know will never get better is a society that is likely to collapse.

The proponents of UBI argue that the payment will prevent us all from starving, and we will supplement our universal basic incomes with activities which we enjoy rather than the wage slave drudgery faced by many people today. But the scenario envisaged here is one in which many or most humans simply cannot get paid for their work, because machines can do it cheaper, better and faster. The humans will still work: they will be painters, athletes, explorers, builders, virtual reality games consultants, and hopefully they will derive enormous satisfaction from it. But they won't get paid for it.

CHAPTER 25

FULLY AUTOMATED LUXURY CAPITALISM

Income

Despite its drawbacks, UBI is at least an attempt to answer the right question, which is: how can we all have a great standard of living, if and when machines take our jobs? In a post-jobs world, there are going to have to be substantial transfers of income and / or wealth from the minority who are still in paid employment, and who own most of the assets. The problem is that if the cost of a good life remains high, then these transfers will be onerous, and therefore avoided. Wealthy people will move to jurisdictions which don't enforce the taxes, or they will simply cease to work at all. And as we saw in the last chapter, we can't fund the transfers by taxing robots, or by eliminating bureaucrats

The solution is not to persecute the wealthy, but to reduce the cost of a good life. This means developing an economy of abundance, in which the prices of all the goods and services that you need for a very good standard of living are very low. Then the transfers from the wealthy need not be onerous – and avoided.

The Star Trek economy

It is often said that science fiction tells us more about the present than it does about the future. Most science fiction writers are not actually trying to predict the future, although they may go to considerable lengths to try to make the worlds they create seem lifelike. In particular, "hard science fiction" writers attempt to make the advanced technology in their stories scientifically plausible. Generally, they are just trying to tell an entertaining story, or maybe use the opportunity that the genre offers to explore something about the fundamental nature of our lives. Science fiction allows writers to conduct philosophical thought experiments: to set new ground rules in a fictional world in order to see what that reveals about the profound questions that we all ask ourselves from time to time. At its best, science fiction is philosophy in fancy dress.

Intentionally or otherwise, science fiction does a very important job for all of us when we think about the future: it provides us with metaphors and scenarios. For better or worse, when prompted to think about artificial intelligence, most people will think about the Terminator, HAL from 2001, or Ava from Ex Machina, for instance. Many of the most popular science fiction stories present dystopian scenarios: think "Blade Runner", "1984", "Brave New World", and so on. But there are also positive scenarios, and one of the most popular ones is "Star Trek".

Set in the 24th century, Star Trek presents a world of immense possibility, of interstellar travel, adventure, and split infinitives. And a world without money or poverty. In the 1996 movie "Star Trek: First Contact", Captain Jean-Luc Picard explains that "Money doesn't exist in the 24th century. The acquisition of wealth is no longer the driving force in our lives. We work to better ourselves and the rest of humanity."

This was not a feature of the original TV series, in which there were quite a few mentions of money, and systems of credit. But before he died, Gene Rodden-berry stipulated that there was to be no money in the Federation.[393]

Although there is no money, the people in the later Star Trek stories do compete with each other – for prestige, for approval, for increased responsibility and for career advancement. One of the things that makes James Tiberius Kirk an outstanding starfleet commander is his fiercely competitive nature. He operates in a profoundly meritocratic environment, and will sacrifice a great deal to win.

This is not new. Men and women have always competed for pre-eminence within their tribes and societies, and we are continually applying our ingenuity to work out new ways to do so. Mediaeval knights risked life and limb for honour and glory, and their descendants fought for national self-determination. Today, many people expend considerable sweat and tears – if less blood – to demonstrate their prowess in sports, in debates, in writing elegant open source software, and in editing Wikipedia pages.

Money is not required in Star Trek's United Federation of Planets because energy has become essentially free, and products can be manufactured in "replicators", devices which create useful (including edible) objects out of whatever matter is available.

Another popular science fiction series with a broadly optimistic (if darkly humorous) outlook is the late Iain M. Banks' "Culture" books, set in a distant future when a technologically advanced humanity has colonised swathes of the galaxy, and enjoys mostly peaceful relations with a host of alien civilisations. The humans are kept company and aided by vastly superior and extraordinarily indulgent machine intelligences, and they lead lives of perpetual indulgence. As Banks put it in a 2012 interview, "It is my vision of what you do when you are in a post-scarcity society, you can completely indulge yourself. The Culture has no unemployment problem, no one has to work, so all work is a form of play."[394]

Abundance

The Star Trek economy is an extreme example of a post-scarcity economy, an economy of abundance. In their 2012 book "Abundance: the future is better than you think", Peter Diamandis and Stephen Kotler argue that this world is within reach in the not-too-distant future, thanks largely to the exponential improvement in technology.

Could they be right? Could the solution to the income problem raised by the economic singularity be that all the goods and services that we need for a rich and fulfilling life (and emphatically not a basic one) are produced so efficiently by machines that they become almost free?

This does sound crazy when you first hear it, but if you think about the music industry, you can see how it could happen. Twenty years ago, not even a rich person could listen to any piece of music which took their fancy. Now it costs $10 a month, thanks to Spotify and similar services. Music, of course, is now digital, so it is dematerialised and non-rivalrous, as economists say. But more and more of what we value these days is digital, and this trend will continue. In another decade or so, many of us will probably spend hours at a time in virtual realities.

Clothify and Constructify

Can we reduce the cost of non-digital goods and services too? Can we create "Clothify" and "Constructify" before nanotechnology gives us replicators? Yes we can - if we do three things.

First, we must take the expensive humans out of the production process for more and more goods and services. That is exactly what automation does, so this is a case where the problem is also part of the cure. We should therefore not hold automation back; instead we should accelerate it. The mantra for companies and for countries should be: "automate and redeploy, rinse and repeat". Companies which automate rapidly will prosper, but the outstanding ones will be those which also genuinely value their people: companies where people actively seek to work out how to automate their tasks, in the knowledge that they will be rewarded both with pay and with more interesting work in the future. Ideally, we should get to a point where people who figure out how to automate their own jobs are regarded as heroes, and rewarded appropriately.

As companies automate routine processes, they will usually have the opportunity – indeed, the obligation - to expand into more sophisticated processes. They can do this by hiring new people with the necessary skills, or by retraining existing staff. A 2019 report on automation by McKinsey claimed that "although re-skilling takes a good deal of effort, it often offers a higher return on investment, in the longer term, than hiring ... On average, replacing an employee can cost 20 to 30 percent of an annual salary, reskilling less than 10 percent."[395]

Second, we must make energy very cheap. At the moment, despite periodic claims by the green lobby that solar power is competitive with fossil fuels, we subsidise the former, and tax the latter. But solar power generation is on a steep downwards cost curve, and our technology for storing and transporting it is improving rapidly. Replacing fossil fuels with electricity generated more directly by the sun isn't going to happen as soon as we might like, but it will happen. Many observers think that within 20 or 30 years, electricity could not only be much cheaper to generate, store and transmit than the oil and gas we get by digging up dead dinosaurs – it could be almost too cheap to meter. This would also reduce CO2 emissions much faster than most people today think possible.

More From Less

Third, we must use AI to make all production processes as efficient as possible, continually reducing the resources used. In his book "More From Less", Andrew McAfee provides compelling evidence that we are already on this path.[396] Any number of environmentalists and lobby groups will tell you that we are polluting, deforesting, and generally destroying the planet, exhausting its natural resources, and driving most other species extinct. All this is making us sick, and crucially, the damage is accelerating.

Implausible as it will seem to many, the data shows the opposite. As we get richer, we are using resources more efficiently, using less energy, causing less pollution and cleaning up the pollution of the past. We are even re-foresting the earth and protecting other species. McAfee produces compelling data and numerous examples, but sadly, many people will refuse to believe him: good news is no news, and if it bleeds, it leads. We do love a good horror story.

Evidence of America's declining resource consumption comes from the US Geological Survey, a federal agency formed in 1879. It tracks seventy-two resources, from aluminium to zinc, and only six of them are not yet post-peak. Even energy usage is decreasing, down two percent in 2017 from its 2008 peak, despite a 15 percent growth in GDP between those two years.

Milk and aluminium are two of McAfee's specific

examples of how America is getting more and more efficient. Between 1950 and 2015, US milk production rose from 117 billion pounds to 209 billion, while the herd shrank from 22 million cows to 9 million. This is a productivity improvement of 330 percent. When aluminium cans were introduced in 1959 they weighed 85 grams. This fell to 21 grams by 1972, and by 2011 it was down to 13 grams.

There is still a long way to go, and transportation provides a good example of the progress we will see in the future. Self-driving technology removes the need for an expensive human driver, and will sharply reduce the number of expensive accidents, along with congestion. Switching from internal combustion engines to electric power reduces the number of moving parts, and removes the wear and tear from combustion. A decade or two from now, solar generated electricity will be significantly cheaper than petrol and diesel, and less polluting too. The cost of getting from A to B could become trivial.

If this post-scarcity economy of abundance is actually possible, can we transition to it without major social upheaval, and possibly collapse? Peter Diamandis talks of building a bridge to abundance, and this may turn out to be the most important challenge for humanity in the first half of this century.[397]

Fully Automated Luxury Capitalism

What form of economic and political governance is most likely to facilitate and sustain this post-jobs world of abundance? Some have argued for a version of communism, called fully automated luxury communism. While this approach deserves credit for facing up to the exponential future, it overlooks two important facts. First, it is no coincidence that wherever communism has been tried, it has degenerated into some of the worst regimes the world has ever seen. Communism grants absolute power to a ruling elite, and we all know what absolute power does to people.

The other fact is the astonishing power of markets. Markets incentivise people and firms to provide goods and services that are genuinely valued, and to do it efficiently. Allied with technology, free markets (with appropriate welfare safety nets, and with regulation to ensure that markets are not suborned) have made this the best time ever to be a human.

Technology gives us new ways to solve old problems, and capitalism provides the incentive for people to invent these new ways and to implement them once they have been invented. Andrew McAfee quotes Abraham Lincoln: "[we add] the fuel of interest [capitalism] to the fire of genius [technology] in the discovery and production of new and useful things." This is how China has gone from backwater to superpower in a single generation.

Even in the abundance economy, it will be a very long time before we have the nanotechnology to build the matter replicators found on Star Trek's USS Enterprise, so goods and services will impose a residual cost. And as long as we are human, we will also face shortages of things like attention, artisanal goods, and works of art. Markets are the best system we know of for resource allocation, so to start with at least, the optimal system for the economy of abundance is likely to be a form of capitalism: let's call it "fully automated luxury capitalism".

The genius of the market economy – the principal reason why it is so effective – is that decisions are taken by the people best qualified to take them. The market enables (indeed obliges) each of us to provide truthful signals about what we do and don't want, what we do and don't value. We buy this car and not that car because we prefer it (given our budgetary constraints) and there is no doubt that we are providing a correct signal because we are spending our own money.

Suppliers are highly incentivised by competition to respond to these signals by offering the best possible goods and services. Although it involves a certain amount of waste, as rivals duplicate facilities to develop and produce their output, the competition between them spurs innovation, and raises quality more effectively than any other system we have tried so far.

By contrast, when decisions are made in a centralised, planned economy, somebody is guessing about what is wanted and needed at every level below them. However good their data collection system, and however well-intentioned they are, they will always be out of date and they will often be just plain wrong. There is also a very good chance that corruption will set in, because it is so easy for that to happen. With apologies to Lord Acton,[398] power corrupts, absolute power corrupts absolutely, and corruption is absolutely central to centralised planning.

There's still money to be made

In fully automated luxury capitalism, people would continue to trade and innovate. Fortunes (and also more modest wealth) would continue to be made by creating art, artisanal goods, and other forms of intellectual property. Even when AIs can create objects more beautiful than the ablest human craft worker, there could well be a premium attached to items which are "made by hand". Judiciously trading scarce assets such as beach-front properties, and original Aston Martin DB5s could remain another source of wealth and income.

But not everyone will be able to or inclined to participate in this reduced commercial world. The three central insights underpinning the idea of fully automated luxury capitalism are (1) that within a generation or two, most humans will not have jobs, (2) that everybody must be wealthy – or at least comfortable rather than poor, and (3) that the taxes levied on wealthy people people and organisations must be affordable. The only way to achieve this is to drive prices down – to evolve the economy of abundance.

Income v wealth

This section has focused on income as opposed to wealth – for a reason. Most people have little wealth, and are therefore dependent on income. The US Federal Reserve estimates that two-thirds of Americans have savings equal to less than three months income.[399] Half of them could not cover an emergency expense of $400 without going into debt. This was aggravated by the recession which began in 2007: the average American family's net worth fell from $136,000 in 2007 to $81,000 in 2013. The situation was revealed starkly by the 2020 Covid pandemic, and governments rushed to provide cash support to their citizens, at breathtaking cost.

Income is the important metric when considering the economic singularity, not wealth. Wealth inequality is far more extreme in today's world than income inequality, both globally and within individual nations. It is also much less significant.

The charity Oxfam created a stir in January 2014 by claiming that the richest 85 people own as much as the poorest 50% of the world.[400] It was repeated so widely and was so helpful in fund-raising that Oxfam has repeated it every year since then. The figure may or may not be correct, but it is highly misleading, and it is disingenuous of Oxfam to keep trotting it out each year. Lack of wealth is not the same as poverty. A young professional in New York living a life of luxury and excess may have no net assets, but it would be perverse to describe her as poorer than a North Korean peasant with no debt and a net worth of a few plastic utensils. Furthermore, if it was possible to eradicate this wealth inequality it would not address poverty. If the richest billionaires gave their wealth to the poorest half of the world, it would amount to a one-off payment of around $500 each.[401]

Nevertheless, if you are one of the lucky minority with substantial net assets, you might be wondering how you will be affected if and when technological unemployment takes hold. Will your house be worth more or less in the new economy? How about your vintage Aston Martin, or your collection of fine wines? Until and unless we move to a completely different kind of economy, it is likely that some of the wealthy people – especially those who control the artificial intelligence which creates most of the added value – will remain wealthy, and perhaps become even more wealthy. Perhaps the prices for Stradivarius violins and prime real estate will continue to rise – for some time at least.

What about the holdings of the much larger number of people in the middle – people who have net assets of a few tens or hundreds of thousands of dollars, perhaps up to as much as a million or two? Unless we switch quickly and smoothly to a very generous form of welfare, it seems likely that the price of assets typically owned by these middle class people, such as suburban houses and mass-produced cars, will slide as their owners try to replace lost income by liquidating their property. This could happen quickly, as people look ahead, see what is coming, and decide to cash in before the slide starts in earnest. Asset prices are notoriously hard to predict because they depend on events which cannot be foreseen, and also upon perceptions about what may happen, and perceptions about those perceptions. This is yet another good reason why we should be thinking seriously about these matters sooner rather than later.[402]

CHAPTER 26

PANIC

Fear

Franklin D Roosevelt was inaugurated as US President in March 1933, in the depth of the Great Depression. His famous comment that "We have nothing to fear but fear itself" was reassuring to his troubled countrymen, and has resonated down the years. In the economic singularity, fear will not be our only problem, but it may well turn out to be our first very serious problem.

Fully autonomous, self-driving vehicles will start to be sold during the coming decade – perhaps in five years, perhaps in ten. Because of the substantial cost saving to the operators of commercial fleets, a great many – perhaps most - of the humans driving taxis, lorries, vans and buses will be laid off during the decade which follows.

Well before this process is complete, though, people will understand that it is happening, and that it is inevitable. Most of us will have a friend, acquaintance or family member who used to be a professional driver. And the technology that destroyed their job will be very evident. One of the interesting and important things about self-driving cars is they are not invisible, like Google Translate, or Facebook's facial recognition AI systems. They are tangible, physical things which cannot be ignored.

Most people are not thinking about the possibility of technological unemployment today. They see reports about it in the media, and they hear some people saying it is coming and others saying it cannot happen. They shrug – perhaps shudder – and get on with their lives. This response will no longer be possible when robots are driving around freely, and human drivers are losing their jobs. This cannot fail to strike people as remarkable. Learning to drive is a difficult process, a rite of passage which humans are only allowed to undertake on public roads when they are virtually adult. The fact that robots can suddenly do it better than humans is not something you can ignore.

No doubt some people will try to dismiss the phenomenon by explaining that driving wasn't evidence of intelligence after all: like chess, it is mere computation. Tesler's Theorem – the definition of AI as whatever we cannot yet do - will cling on. But most people will not be fooled. Self-driving vehicles will probably be the canary in the coal mine, making it impossible to ignore the impact of cognitive automation. People will realise that machines have indeed become highly competent, and they will realise that their own job may also be vulnerable.

If we have a Franklin D. Roosevelt in charge at the time – perhaps one in every country – this may not be a problem. If there is a plausible plan for how to navigate the economic singularity, and a safe pair of hands to implement the plan, then we may be OK. Unfortunately we do not currently have a plan. There is no consensus about what kind of economy could cope with a majority of the population being permanently unemployed, nor how to get from here to there. Neither are all the top jobs in safe pairs of hands.

In the absence of a solid plan explained by a reassuringly competent leadership, the reaction of large numbers of people realising that their livelihoods are in jeopardy is not hard to predict: there will be panic.

When will this panic occur? Within a few years, and perhaps within a few months, of self-driving vehicles starting to lay off human drivers. In other words, in a decade or so.

Populism

Populist is a title applied to politicians who claim to represent the common people in a perceived battle against the interests of an established elite. Few people actively claim the title, as it also denotes a politician who promises much but will deliver little, and will probably cause great harm to those who elected him or her, as well as to everyone else.

President Trump and the British campaign to leave the EU certainly claimed to champion the common people against an established elite. You will have your own view about whether the rest of the description applies to them. It is usually thought that the success of these campaigns was due to a feeling among voters that they have been "left behind" and economically disadvantaged since the great recession of 2008 onwards. In which case it is odd that populists had their greatest successes in countries which performed better economically than most others. US GDP was 0.26% higher in 2016 than in 2008, which is hardly a sparkling performance, but it is considerably better than the EU, whose GDP shrank by 0.14%.[403] The UK's GDP grew by 0.8%, just a whisker behind the most successful large European economy, Germany, whose GDP grew by 0.9%.

Why did the relatively successful USA and UK experience extreme populism, when the less successful France (down 0.16%), Italy (down 0.23%) and Spain (down 0.25%) did not?

There is a clue in the most prominent slogan of each campaign: "Make America Great Again", and "Take Back Control". Trump and Brexit were backward-looking, nostalgic campaigns which promised a return to a better time that supposedly existed in the past. Yes, they were campaigns against allegedly corrupt and smug elites (Wall Street and Washington in the US, "the metropolitan elite" in the UK, and the national broadcast media in both), but first and foremost they were campaigns against change.

The change they were seeking to reverse is more cultural than economic. Trump and Brexit were a reaction against the decades-long triumphal march of social liberalism, which overturned what many people believed to be the natural order of things. It is no coincidence that the formation of the Tea Party, the start of the current wave of right-wing populism in the US, was in 2009, the year of Obama's inauguration. The election of a black president was perhaps the apogee of liberal values.

For many Trump and Brexit voters, the single most important issue was immigration. As the West's most open economies, and among their most successful, the US and the UK attracted more than their fair share of young, ambitious expatriates since the 2008 crash. In the early 2010s, if you were young, entrepreneurial and Polish, Greek, Mexican or Vietnamese, you would be quite likely to decide that your future was more promising in London, New York or San Francisco than in your home country. Once there, if not before, you would probably adopt a left-leaning, politically correct world-view, and this, together with the simple fact of being foreign, would make you an object of suspicion in the eyes of a native population that was less open to change, and feared your impact on its employment prospects and access to public services.

Panic

The election of Trump and the Brexit referendum result were political earthquakes. Politics has not been so "interesting" since at least the fall of the Berlin Wall and the end of the Cold War at the end of the 1980s. But compared with a realisation by the majority of the population that they are very likely to lose their jobs, their causes were relatively minor. The possible impacts of a panic about impending widespread joblessness could be enormous, and they are worth expending considerable effort to avoid.

In chapters 28 and 29 we will explore how all this might unfold in practice. But first, we will try to peer beyond the event horizon of the economic singularity, at two of the longer-term challenges it raises.

CHAPTER 27

BEYOND THE ECONOMIC SINGULARITY

Event horizon

The canonical example of a singularity is a black hole. As we saw in chapter 15, at the centre of a black hole the gravitational field becomes infinite, and the laws of physics stop working – or at least, they change fundamentally. This is why John von Neumann used the term "singularity" to denote the enormity of the social and economic change which accelerating technology would cause.

Another characteristic of a black hole is the event horizon, a boundary beyond which events cannot affect an observer. A black hole has an event horizon because light cannot escape its gravitational field, and cannot reach an external observer.

(Amazingly, the idea that the universe might contain objects so massive that light could not escape them was first suggested in 1784, by a remarkably prescient English clergyman and natural philosopher called John Michel. And Stephen Hawking's discovery that radiation leaks from black holes means that "gravitational collapse produces apparent horizons but no event horizons".[404] But we are using a metaphor here, not studying astrophysics.)

An obvious feature of an event horizon is that it is extremely hard to determine what happens beyond it. But let's try. This chapter outlines two problems which may be raised by a successful transition through the economic singularity: the allocation problem, and the speciation problem. It also offers rather hazy solutions to both.

The Allocation Problem

Imagine we manage to cross Peter Diamandis' bridge to an economy of abundance. We are all living comfortable and fulfilling lives. Most of us take advantage of the almost-free goods and services which are provided by intelligent machines, and the costs are met by (non-punitive) taxes on the incomes of those who still have incomes, and on the assets owned by the talented and the lucky.

Not everything can be free or nearly free, and many items will always be scarce. There is a finite and regrettably small supply of large houses on empty white sand beaches fringed with palm trees leading down to a turquoise sea. Or penthouse apartments on Manhattan's Fifth Avenue. There is a very small supply of Vermeers and Aston Martin DB5s.

How will we allocate those goods and services which cannot be rendered free or nearly free? The answer seems simple and obvious: we will still have money, and we will still have the market. Supply and demand will continue to operate like before.

But most people will have very little money, and hence very little opportunity to acquire the rare and expensive items. And well below the level of Fifth Avenue apartments, the quality of the assets that we use and enjoy will be extremely varied. In every village, town and city, there is a wide variety of quality of housing stock.

The society of abundance will generate egalitarian incomes for the large numbers of people who are not employed, but a decidedly un-egalitarian asset base. Everyone's access to new goods and services will be more-or-less the same, at least within a given territory or jurisdiction, but some people will be living in nice big houses in the posh part of town, while others will be living in small flats with no sound-proofing in grubby apartment blocks in the unfashionable zone.

Will it be like a game of musical chairs? Prior to the economic singularity we all work hard to improve our lot, and then when the machines take our jobs, the music stops and we all sit down in the chairs we have arrived at? And just stay there forever? That seems neither fair nor sustainable.

Do we decide that no-one can own the precious things? Perhaps we could turn all the nice houses into museums and keep the scarce movable objects on display there, to be visited (and perhaps used) on payment of a fee, or by scheduled appointment.

In the meantime we set the efficient machines to revitalising and / or replacing the stock of lower-quality houses. (And cars, and boats, and furniture, and clothes, etc.) But it will take a very long time indeed to build a nice new house for everyone who doesn't start off with one. And even when we have completed that gargantuan task, some houses will still be in much nicer places than others.

And who will decide what the cut-off point is between a house which people can carry on living in, and one which is too nice to be private property?

This is the allocation problem.

VR to the rescue?

In February 2016, when Palmer Luckey and John Carmack were the key executives of Oculus VR, a Facebook-owned manufacturer of VR hardware and software, they talked about a "moral imperative" to make virtual reality available to us all.[405]

"Everyone wants to have a happy life, but it's going to be impossible to give everyone everything they want.... Virtual reality can make it so anyone, anywhere can have these experiences." "You could imagine almost everyone in the world owning [good VR equipment]. ... This means that some fraction of the desirable experiences of the wealthy can be synthesized and replicated for a much broader range of people."

Other people have thought about these questions, and not everyone is delighted by the suggestion that VR can assuage the frustration caused by scarcity.

Some people think it impossible, and others think it possible but degrading.

The Harvard political philosopher Robert Nozick described a thought experiment back in 1974 featuring an "experience machine" which could recreate any sensation you choose. Your brain is persuaded that the experience is real, which means that you believe it too, but in fact your body is lying in in a flotation tank, deprived of all sensory input while your brain is hooked up to the machine. Philosophers do a lot of their work by investigating their intuitions, and Nozick's intuition was that no-one would use this machine because we value reality too highly. I find it surprising that he came to that conclusion back in 1974, and it would be an even more surprising conclusion to reach today, when so many people spend so much of their lives in simulated realities, albeit only imperfectly simulated. Certainly a great deal of money is being invested by smart people in the belief that we will consume VR avidly. Nozick died in 2002, so he won't have to find out for himself – maybe he would be relieved.

Other critics see the Oculus founders' view of the future as possible but frightening. Ethan Zuckerman is director of the MIT Centre for Civic Media, and thinks that "the idea that we can make gross economic inequalities less relevant by giving [poor people] virtual bread and circuses is diabolical and delusional." Jaron Lanier is a computer scientist and writer who founded VR pioneer VPL Research, and is generally credited with popularising the term virtual reality. He lambasts as "evil" the vision that the rich will become immortal, while "everyone else will get a simulated reality. ... I'd prefer to see a world where everyone is a first-class citizen and we don't have people living in the Matrix."

Only time will tell if VR is helpful, or even necessary, in enabling us to live in a world where machines have made humans unemployable. Clearly it will play a major role in the lives of most people, and it should make those lives more productive, more fun and more fulfilling. As Oculus' John Carmack puts it, "if people are having a virtually happy life, they are having a happy life. Period."

The allocation problem is severe, and potentially destabilising for society. The speciation problem is even more so.

"The Gods and the Useless"

In his remarkable book "Homo Deus", Yuval Harari makes the brutal suggestion that sooner or later, most people will be unemployable, and the consequence will be a two-tier economy consisting of the gods and the useless. "As algorithms push humans out of the job market, wealth might become concentrated in the hands of the tiny elite that owns the all-powerful algorithms, creating unprecedented social inequality. ... The most important question in twenty-first-century economics may well be what to do with all the superfluous people."

Imagine a society where the great majority of people lead lives of leisure, their income provided by a beneficent state, or perhaps a gigantic charitable organisation. They are not rich, they don't travel first class or frequent expensive restaurants, and they don't own multiple houses. But they have no pressing needs and in fact they want for little: they enjoy socialising, learning, sports, exploration, and much of this is carried out in virtual worlds which are almost indistinguishable from reality.

A small minority of people in this society do have jobs. Their work is pleasurable and intellectually simulating, and not stressful. It involves monitoring and occasionally guiding or re-setting the performance of the machines which run their society – machines which they own.

Let's say that this elite minority is generous towards the majority which lives outside their gated communities, and which does not visit the luxurious resorts they migrate between, and does not travel with them on their private heli-jets. They are effectively benign rulers, although both camps refrain from putting it like that.

In this future world, all members of the species of homo sapiens are changing. They are using new technologies to enhance themselves both cognitively and physically. They use smart drugs, exoskeletons and genetic technologies, among others. Maybe they have engineered themselves to need less sleep.[406]

Everyone has access to these technologies, but the elite has privileged access. They get them sooner, and this could be vitally important. Historically, concerns about a "digital divide" were exaggerated. Companies make much more money by selling lots of relatively cheap cars and smartphones to almost everyone than they ever could by selling just a handful of diamond-encrusted versions to the super-rich. Effectively, the rich are guinea pigs for new technologies, trying out primitive versions of new gadgets which don't work very well, and enabling producers to start climbing the learning curve to manufacture them at scale, which in time enables almost everybody to have them.

Speciation

But technology is advancing at an accelerating rate. In the future society we are envisioning, important breakthroughs in physical and cognitive enhancement are announced every year, then every month, then every week. As artificial intelligence gets better and better, it fuels this improvement – even though it is still narrow AI, and still far from human-level, artificial general intelligence, or AGI.

It may become hard or even impossible to disseminate these cognitive and physical improvements quickly enough to avoid a profound separation between those with privileged access to them and the rest of us.

So the elite will change faster than the rest. As the two groups lead largely separate lives, the widening gap may not be apparent to the majority, but the elite surely will know about it. They will decide that they must draw attention away from the fact, and they will take precautions to prevent attack, in case the majority should become aware of and resentful about what is happening. They will surround their gated communities with discreet machines which possess astonishingly powerful defensive – and offensive – capabilities. They will keep themselves more and more to themselves, meeting members of the majority less and less often. When they do meet, it will almost always be in virtual reality, where their avatars (their representations in the VR environment) do not betray the widening gulf between the two types of humans.

When normal people read about the lives of billionaires and movie stars, we often think they live in a different world. But the distance which separates them from us is tiny compared to the gulf which could open up between the AI-owning elite and the unemployed majority in a world which passed through the economic singularity while retaining private property.

After a while, humans may speciate: the gods and the useless could actually evolve into two different species.

Brave New World

This is not your average science fiction dystopia. The scenario most commonly offered up by Hollywood is that technology has pulled back the curtain which concealed the truly evil nature of the capitalism system, and mankind has fallen into a form of high-tech slavery, where the rich descendants of company CEOs and scheming tycoons and financiers brutalise an impoverished and oppressed majority.

The movie "Elysium" is just one of many dreary examples of this tired old cliché. In it, as in so many others, society has actually regressed from capitalism into a sort of techno-feudalism. Any viewer who is half-awake is wondering, since machines can do all the work, what is the point of enslaving humans?

(It is curious that the left-leaning culture prevalent in Hollywood impels it to issue these tirades against capitalism, when Hollywood studios are themselves formidable exponents of the capitalist arts, and Hollywood stars earn millions of dollars per film.)

Aldous Huxley's 1931 book, "Brave New World" presents a far more interesting scenario, and a very subtle piece of world-building. Almost everyone is content in the society which has been developed after an appalling military conflict, but it is clear to the reader that humanity has lost something important, and most of it has regressed to an almost infantile condition. Yet when a talented outsider arrives, he is unable to devise a way to improve the system, or to accommodate himself to it. He instinctively feels – and the reader is encouraged to agree – that his own life has shown him there is a better way to live, but he is unable to articulate or maintain it.

In Huxley's story, humanity has achieved a stable equilibrium. Only a tiny minority of humans are aware of what has been sacrificed in order to achieve a society of docility and acquiescence. Regular sex and a powerful drug called soma are the opiates of the masses. (Huxley wrote the book long before the advent of rock and roll, so he couldn't include the third leg of Ian Dury's hedonistic mantra that sex and drugs and rock and roll were all his brain and body needed.[407])

Brave New World is certainly not intended as a blueprint. Even so, it is calm, and the spectre of economic and social collapse seems to have been abolished – at least for the time being. It would be foolish of us to take even that much for granted. A society where the gods and the useless are becoming different species might turn out to be inherently unstable. In a full-on conflict between them it seems likely that the gods would have the means to protect themselves, but at what cost?

Collective ownership

So perhaps navigating the economic singularity successfully requires the asset-owning elites to transfer their assets into collective ownership, and be hailed as heroes and heroines for doing so. How might this work in practice?

I argued above that planned economies have inherent flaws, and that part of the genius of the market economy is that decisions are taken by the people best qualified to take them. Common ownership can work well in small communities, such as families, tribes, and small villages. But as soon as a society attains any level of size and sophistication, the bonds of kinship weaken and individuals start to claim ownership over land and property. The society becomes regulated by power structures which begin as means of self-defence and evolve into expressions of ambition.

If (and it is a big "if") surviving the economic singularity and avoiding fracture means ending the system of private ownership, how can this be done without falling into the unwelcome embrace of an over-mighty state and centralised planning?

The answer just might be the blockchain.

Blockchain

Initial Coin Offerings (ICOs) are a way to raise money for digital currencies and other enterprises employing blockchain technology. In June 2017, the amount of money raised in ICOs overtook the amount invested by the venture capital industry.[408] Given that most people are very hazy about blockchain technology, and that regulators are nervous that ICOs enable Ponzi schemes and money laundering on a huge scale, this was a remarkable milestone. Since then, the brouhaha has receded, but there are still plenty of smart people claiming that blockchain technology is a revolutionary force.

The biggest and most famous application of blockchain technology is bitcoin, which launched as open-source software in 2009, having been described in a paper published in October 2008 under the pseudonym Satoshi Nakamoto. Its valuation has been extremely volatile. At launch a bitcoin was worth virtually nothing, but by March 2017 it reached $1,268, overtaking the value of an ounce of gold for the first time. In December 2017 it reached a peak of $19,783, but a year later it fell below $3,200.[409] In June 2020 it was $9,350.

The blockchain is a public ledger which records transactions. The important thing is that the ledger is completely trustworthy despite having no central authority, like a bank, to validate it. It is trustworthy in that you can have full confidence that if someone gives you a bitcoin, then you do own that bitcoin: the person who gave it to you will not be nipping off to spend the same piece of currency elsewhere, even though it is entirely digital.

This confidence arises because transactions are recorded in blocks which are added to the chain by people (or rather computer algorithms) called miners. These miners are working continuously on mathematical problems whose solutions are hard to find but easy to verify. A problem is solved ("mined") every few minutes, and each solution creates a block. The new block is added to the chain, and incorporates the transactions made since the last block was added to the chain. Your transaction is published on the blockchain's network as soon as it is agreed, but it is only confirmed, and hence reliable, when a miner has incorporated it into a block.

Satoshi Nakamoto's innovation solved a previously intractable challenge in computer science known as the Byzantine Generals' Problem. Imagine a mediaeval city surrounded by a dozen armies, each led by a powerful general. If the armies mount a co-ordinated attack, their victory is assured, but they can only communicate by messengers on horseback who visit the generals one by one, and some of the generals are untrustworthy. The blockchain provides a way for each general to know that a message calling for an attack at a particular time is genuine, and has not been fabricated by a dishonest general before it reached him.[410]

Digital currency is only one of the possible applications of blockchain technology. It can register and validate all sorts of transactions and relationships. For instance, it could be used to manage the sale, lease or hire of a car. When you take possession of a car, it could be tagged with a cryptographic signature, which would mean that you are the only person who could open and start the car.[411]

The revolutionary benefit of the blockchain is that all kinds of agreements can be validated without setting up a centralised institution to do so. By removing the need for a central intermediary, the blockchain can reduce transaction costs, and it can enhance privacy: no government agents need have access to your data without your permission.

Most importantly, for our present purposes, the blockchain may make possible the decentralised ownership and management of collective assets.

The elite's dilemma

Imagine a future in which it is apparent to many people that we are heading towards the scenario of the gods and the useless, and staring right into the face of the speciation problem. The elite few who own the machines are as uncomfortable as the rest of us about this – or at least a sizeable number of them are. They do not want to hand their assets over to a government organisation, as they believe this would simply swap one potentially dangerous elite for another one.

But they fear that if the scenarios of the gods and the useless becomes reality, they will end up as pariahs, feared and perhaps hated by the rest of humanity. This outcome might well be grim for the useless, but it would be unpleasant for the gods as well.

Rich people are not all bad, greedy and selfish. They are the same mix of good and bad, greedy and generous as the rest of us. They tend to be smart and hard-working, but otherwise they are pretty normal, which is to say, that curious human blend of similar and different, happy and sad, predictable and unpredictable.

It is quite plausible that in the gods and the useless scenario, the smallish minority which owns most of the assets when the game of musical chairs stopped – including notably the AI – would prefer to throw in their lot with the rest of us rather than hide behind heavily fortified gates, outcasts from the rest of humanity.

It would be a non-trivial project to work out in detail how the assets could be transferred into universal common ownership, validated by the blockchain, and managed in a decentralised fashion. And it is certainly not a forgone conclusion that the rich minority would endorse it. But it may turn out to be our best way forward.

We will not have to address the allocation and the speciation problems unless we successfully tackle the income problem and transition through the economic singularity. The next two chapters look at how that might work out in practice.

CHAPTER 28

FOUR SCENARIOS

Scenario 1 - No Change

In a July 2015 interview with Edge, an online magazine, Pulitzer Prize-winning veteran New York Times journalist John Markoff lamented the deceleration of technological progress - in fact he claimed that it has come to a halt.[412] He claimed that Moore's Law stopped reducing the price of computer components in 2013, and pointed to the disappointing performance of the robots entered into DARPA's 2015 Robotics Challenge.

He argued that there has been no profound technological innovation since the invention of the smartphone in 2007, and complained that basic science research has essentially died, with no modern equivalent of Xerox's Palo Alto Research Centre (PARC), which was responsible for many of the fundamental features of computers which we take for granted today, like graphical user interfaces (GUIs) and indeed the PC.

Markoff grew up in Silicon Valley and began writing about the internet in the 1970s. He fears that the spirit of innovation and enterprise has gone out of the place, and bemoans the absence of technologists or entrepreneurs today with the stature of past greats like Doug Engelbart (inventor of the computer mouse and much more), Bill Gates and Steve Jobs. He argues that today's entrepreneurs are mere copycats, trying to peddle the next "Uber for X".

He admits that the pace of technological development might pick up again, perhaps thanks to research into meta-materials, whose structure absorbs, bends or enhances electromagnetic waves in exotic ways. He is dismissive of artificial intelligence because it has not yet produced a conscious mind, but he thinks that augmented reality might turn out to be a new platform for innovation, just as the smartphone did a decade ago. But in conclusion he believes that "2045... is going to look more like it looks today than you think."

It is tempting to think that Markoff was to some extent playing to the gallery, wallowing self-indulgently in sexagenarian nostalgia about the passing of old glories. His critique blithely ignores the arrival of deep learning, social media and much else, and dismisses the basic research that goes on at the tech giants and at universities around the world.

Nevertheless, Markoff does articulate a fairly widespread point of view. Many people believe that the industrial revolution had a far greater impact on everyday life than anything produced by the information revolution. Before the arrival of railroads and then cars, most people never travelled outside their town or village, much less to a foreign country. Before the arrival of electricity and central heating, human activity was governed by the sun: even if you were privileged enough to be able to read, it was expensive and tedious to do so by candlelight, and everything slowed down during the cold of the winter months.

But it is facile to ignore the revolutions brought about by the information age. Television and the internet have shown us how people live all around the world, and thanks to Google and Wikipedia, etc., we now have something close to omniscience. We have machines which rival us in their ability to read, recognise images, and process natural language. And the thing to remember is that the information revolution is very young. What is coming will make the industrial revolution, profound as it was, seem pale by comparison.

The productivity paradox

Part of the difficulty here is that there is a serious problem with economists' measurement of productivity. The Nobel laureate economist Robert Solow famously remarked in 1987 that "you can see the computer age everywhere but in the productivity statistics." Economists complain that productivity has stagnated in recent decades. Another eminent economist, Robert Gordon, argues in his 2016 book "The Rise and Fall of American Growth" that productivity growth was high between 1920 and 1970 and nothing much has happened since then.

Anyone who was alive in the 1970s knows this is nonsense. Back then, cars broke down all the time, and were also unsafe, and highly polluting. Television was still often black and white, it was broadcast on a tiny number of channels, and it was shut down completely for many hours a day. Foreign travel was very expensive, and hence rare for most people. And we didn't have the omniscience of the internet. Many of the dramatic improvements to this pretty appalling state of affairs are simply not captured in the productivity or GDP statistics.

Measuring these things has always been a problem. A divorce lawyer deliberately aggravating the animosity between her clients because it will boost her fees is contributing to GDP because she gets paid, but she is only detracting from the sum of human happiness. The Encyclopedia Britannica contributed to GDP, but Wikipedia does not. The computer you use today probably costs around the same as the one you used a decade ago, and thus contributes the same to GDP, even though today's version is a marvel compared to the older one. GDP completely misses environmental harms and reparations, and changes in healthcare, education, and freedoms. It seems that GDP is becoming increasingly divorced from the things that humans care about. It may well be that automation will deepen and accelerate this phenomenon.

The particulars of the future are always unknown, and all predictions are perilous. But the idea that the world will be largely unchanged three decades hence seems the least plausible of the scenarios set out in this chapter.

Scenario 2 - Full employment

In chapter 13 we reviewed two arguments against the thesis that cognitive automation will lead to lasting widespread unemployment, and seven examples of ways in which humans can remain employed as robots take our old jobs.

The arguments were what I characterise as the reverse Luddite fallacy, and the argument from inexhaustible demand. The reverse Luddite fallacy says that automation has not caused lasting unemployment in the past, and therefore it cannot in the future. When stated as baldly as that it is clear how weak the argument is: past performance is no guarantee of future outcome, and in any case, automation did cause massive unemployment in the past – of horses.

The argument from inexhaustible demand does not explain how humans can satisfy any of the demand when machines can do most things cheaper, better and faster. If humans and machines chase each other down the revenue curve in satisfying demand, humans will reach the point where it is not worth them continuing long before machines do.

But the fact that the arguments purporting to show that technological unemployment cannot happen are weak does not mean that it will necessarily happen. The fact is that we do not yet know for certain. The seven examples of how humans could perhaps keep working can be summarised as follows.

(i) We will form centaur-like partnerships with machines in which they will take care of brute force calculations and lower-level cognitive tasks, and humans will provide the imagination, creativity and flair.

(ii) The ability to trawl through masses of data and surface all possible correlations will make it possible to carry out tasks which were previously impossible, or were previously unaffordable. We saw this in the case of lawyers becoming able to review thousands of employment records at a much lower price point, and it is called the iceberg effect.

(iii) Humans will do jobs which require empathy, which machines cannot have because they are not conscious.

(iv) Humans will buy products (and perhaps services) from other humans in preference over goods made by machines because of an innate chauvinism, or because the human-produced goods have the artisanal quality of being imperfect and slightly different, one from the next.

(v) Humans will be entrepreneurs, which machines cannot because they have no ambition.

(vi) Humans will be artists, which machines cannot because art requires the communication of an experience, and unconscious machines can have no experiences.

(vii) If all else fails, we will open the magic jobs drawer, and out will fly all kinds of new activities that we cannot imagine today because the technology to make them possible has not yet been invented.

As we saw, there are strong counter-arguments and counter-examples to each of these, and it seems unlikely they can keep us all in employment. But collectively, they provide comfort to the technological unemployment skeptics, and the self-professed "deniers", like Google's Eric Schmidt.

Scenario 3 – Dystopia

Civilisation is fragile. Any schoolchild can name some great empires which collapsed: the Greeks, the Persians, the Romans, the Maya, the Inca, the Mughals, the Khmer, the Ottomans, the Hapsburgs. The ancient Egyptians managed to rise and fall several times during their extraordinary 3,000-year history.

We also know how fragile civilisation is from two famous episodes in experimental psychology. In 1961, Yale psychologist Stanley Milgram recruited students from that elite university, and told them to administer mild electric shocks to incentivise a stranger who was supposed to be learning pairs of words. The shocks were fake, but the students did not know this, and an extraordinary two-thirds of the students were prepared, when urged on by the experimenter, to deliver what appeared to be very painful and damaging doses of electricity.[413] The experiment has been replicated numerous times around the world, with similar results.

Ten years later, Stanford psychology professor Philip Zimbardo, a school friend of Milgram's, ran a different experiment in which students were recruited and arbitrarily assigned the roles of prisoners and guards in a make-believe prison. He was shocked to see how enthusiastically sadistic the students who were chosen to be guards became, and he was obliged to terminate the exercise early.[414] This experiment has also been replicated numerous times.

Our 21st century global civilisation seems pretty robust. We have just gone through what is frequently described as the worst recession since the Great Depression of the 1930s, and for the great majority of people, the experience was nothing like as awful as those terrible years, which did so much to set up the disastrous carnage of World War Two. It is too soon to say how bad the aftermath of the Covid-19 pandemic is going to be.

But history and experimental psychology demonstrate that we cannot afford to be complacent. If the argument of part three of this book is correct, we are about to embark on a journey towards a new type of economy which we have not yet designed. Unless we are careful, there will be plenty of opportunities for mis-steps, misunderstandings, and downright mischief by populists and demagogues.

If technological unemployment arrives in a rush, and we are not prepared, a lot of people will lose their incomes quickly, and governments may not move fast enough to avert drastic collapses in asset prices as people sell their belongings to make ends meet. If the replacement of incomes is slow or botched in some countries, the resulting economic crises could lead to their governments being overthrown by irresponsible or foolish leaders. Among other things, we must hope this does not happen in any countries with significant stocks of nuclear weapons.

In the developed countries, and increasingly, elsewhere too, our lives are intertwined and inter-dependent. Especially if we live in cities – which more than half of us now do – we depend on just-in-time logistics systems to deliver food and other essentials to our local shops. How long would any of us survive if all the supermarkets suddenly lost their supply chains, and everyone around us was getting hungry?

Preppers

There are signs that the people best-placed to understand what is coming have a keen appreciation that the changes bearing down on us may not end well. Reid Hoffman, co-founder of LinkedIn, told a journalist in early 2017 that half his peer group of Silicon Valley billionaires have some level of "apocalypse insurance" in the form of a hideaway in the US or abroad.[415] New Zealand is a favourite location, as what its natives used to call the "tyranny of distance" becomes a virtue if the rest of the world is going to hell in a handcart.

"Saying you're 'buying a house in New Zealand' is kind of a wink, wink, say no more. Once you've done the Masonic handshake, they'll be, like, 'Oh, you know, I have a broker who sells old ICBM silos, and they're nuclear-hardened, and they kind of look like they would be interesting to live in.' "

Preppers – people who are preparing for the worst - have their own language. They are preparing for the day when SHTF (the Shit Hits The Fan) and the USA is WROL (Without the Rule Of Law). They joke that FEMA, the Federal Emergency Management Agency, actually stands for Foolishly Expecting Meaningful

Aid.

It would be ironic if the small percentage of people who survived a massive, widespread social collapse included many of those who invented and deployed the technologies that helped to cause it.

Populism to fascism

Collapse is not the only dystopian outcome. Self-driving vehicles may well bring with them an understanding that widespread lasting unemployment is coming, and perhaps a consequent panic. In some countries at least, that may bring forth strong leaders who promise security, and law and order. In these countries, society may peer over the brink and decide collectively to step back and away from collapse. They may accept a trade-off in which they surrender much of their liberty, and perhaps many of the rights they used to take for granted, in exchange for a guarantee of some kind of rule of law.

Seizing control of the media, of social media, and making full use of all the cameras and sensors in the environment, a determined government could probably exercise a formidable degree of control, and could perhaps cow and crush any resistance. In the absence of a genuine solution to the challenges posed by technological unemployment, various bogus solutions would be tried out, and their abject failure would be denied and covered up. The result would almost inevitably be economic reversal and hardship, but again, this might be an acceptable alternative to collapse.

The populist leader would demand slavish obedience, and the outward expression of loyalty. "Out" groups would be demonised because of their race, religion, sexuality, or political beliefs. Fear and hatred would be incited against an external enemy, a rival nation or group of nations, which required the domestic population to unite and make sacrifices to defeat.

We have seen this story too many times before, in fascist regimes of the right and the left. It always creates misery, and it usually leads to war. So in the long run it may not be an alternative to collapse at all.

Fracture and then collapse

Another dystopian vision which could be realised some years further into the future is speciation: the gods and the useless scenario that we explored in chapter 27. It could be a stable if unappetising arrangement, like the world envisioned in Huxley's Brave New World. But the encounters between the Spanish conquistadores and the Aztecs in Mexico and the Inca in Peru are two of many which suggest that encounters between two human civilisations do not work out well for the one which is less technologically advanced. It seems a reasonable expectation that sooner or later the gods and the useless scenario would lead to serious trouble for one or both parties.

Scenario 4 - Protopia

Kevin Kelly is a writer, and the founding editor of Wired magazine. He has been called the most interesting man in the world.[416] I have no idea whether he enjoys the burden of that appellation, but he does produce a lot of interesting ideas. One of his good ones is Protopia.

Too much of today's thinking about the future is dystopian, and that is partly because too many people fail to realise just how much progress homo sapiens has made in the last few centuries and decades. It is natural and indeed helpful for our species to be discontented: if we weren't discontented, we probably wouldn't struggle to make the world a better place. But it can lead to dangerous misunderstandings.

Many people think that all politicians are corrupt, and that all corporations are run by Bond villains who are greedy and bent on world domination. Most of us could think of some group, clique or tribe that we are suspicious, fearful or disdainful of. But the truth is that in most of the world, today is the best time there has ever been to be alive. Most people in developed countries today live better than kings and queens did a couple of centuries ago. We live longer, eat better, have better healthcare, and inconceivably better access to information and entertainment than previous generations. If you doubt this, take a look at one or two of the late Hans Rosling's delightful and inspiring TED talks,[417] and browse the charts at "Our World in Data" compiled by Oxford University researcher Max Roser.[418]

Of course everything could go horribly wrong tomorrow. There might even be an iron law of nature that when civilisations reach a certain stage they either blow themselves up, or create machines to do it for them. But from where we stand today, there is no reason to believe that. It seems more likely that the future is open, and potentially very good indeed.

Utopian visions of the future are less common, but they are also problematic. A future in which life has to all intents and purposes become perfect sounds sterile and boring. It is also highly improbable: the more we learn about the universe, the more we discover that we don't know, so it seems unlikely the universe will one day stop presenting us with puzzles and challenges. Perhaps this is why the two best-known literary descriptions of utopias, Thomas More's "Utopia" and Voltaire's "Candide", are essentially critiques of the societies they lived in rather than recipes for an ideal future one.

Interestingly, the word "optimism" now means the opposite of what it started out meaning in the 18th century, which was the belief that life was already optimal, and therefore could not improve.

So it is refreshing to read Kelly saying this: "I am a protopian, not a utopian. I believe in progress in an incremental way where every year it's better than the year before but not by very much—just a micro amount. I don't believe in utopia where there's any kind of a world without problems brought on by technology. Every new technology creates almost as many problems that it solves." But crucially, it gives us "a choice that we did not have before, and that tips it very, very slightly in the category of the sum of good."[419]

The good life

That "sum of good" might be what the ancient Greeks

called "eudaimonia": the good life that we all seek (or at least, should seek), in which we humans flourish. The Greeks debated vigorously whether eudaimonia consisted of happiness, or virtue, or both. Their concept of virtue included being good at something, as well as the moral virtues lauded by Christianity and other monotheistic faiths. At the risk of over-simplifying, Socrates, and later on the Stoics, thought that eudaimonia required the exercise of virtue alone, because virtue is both necessary and sufficient for happiness. Aristotle disagreed, saying that eudaimonia requires both happiness and virtue. A completely virtuous person would not enjoy eudaimonia if, for instance, all her children died. Epicurus added another twist, arguing that happiness was the only necessary ingredient of eudaimonia, but that happiness is impossible without leading a life that is also virtuous. Confused? Welcome to the human condition.

The first two scenarios explored in this chapter were perfectly acceptable, but unrealistic. The third was dystopian. What would the fourth scenario – protopia - look like? It might unfold in four stages: a plan, big welfare, Star Trek, and collectivism. We will flesh these stages out in the next chapter.

CHAPTER 29

PROTOPIAN UN-FORECASTS

Three snapshots of a positive scenario

This chapter offers three snapshots of a possible future, taken in 2030, 2040 and 2050.[420] Their purpose is to make the possibility of technological automation seem more real and less academic, and to explore how a positive outcome might unfold, leading to an economy of abundance without going through massively damaging social dislocation. Each snapshot describes the impact of cognitive automation on society as a whole at that time, and also on the main sectors of the economy.

Unpredictable yet inevitable

We know that all forecasts are wrong. The only things we don't know are by how much, and in what direction. The future generally turns out to be not only different to what we expect, but also much stranger. Cast your mind back to 2005. Pretty much everyone thought that cellphones would continue to get smaller, and Facebook was limited to a few thousand universities and schools. Today, just a decade later, larger smartphones are a bit of a thing, and Facebook's valuation has overtaken that of Walmart, the world's largest shopkeeper.[421] Trying to predict how the world will look in 2030 is like trying to predict the weather on Saturday two months from now. There are just too many variables.

And yet in hindsight what happens appears not only natural, but almost inevitable.

The smartphone is a good example. Pretty much nobody suggested thirty years ago that we would all have telephones in our pockets which would contain powerful artificial intelligences, and which would only occasionally be used for making phone calls. After all, at the time a mobile phone was a fairly hefty device, the size of a small dog. But now that it has happened it seems obvious, logical, and perhaps even inevitable.

Here's why. We humans are highly social animals, and our social habits are facilitated by language. Because we have language we can communicate complicated ideas, suggestions and instructions: we can work together in teams and organise; we can defend ourselves against lions and hostile tribes, we can hunt and kill mammoths, produce economic surpluses, and develop technologies.

It is often said that no species is more savage and more violent than humans. This is no more true than the claim that Americans are more violent than other nationalities because their murder rate is higher than other developed countries. America's murder rate is high because its citizens can buy guns so easily. Humans kill more than other species because we have more and better weapons.

Humans live cheek by jowl in cities containing millions. This is remarkable: no other carnivorous species can assemble more than a few dozen of its members in a limited space without them killing each other in rivalry for food, sex, or social dominance. Other species lack our sophisticated ways to communicate and collaborate. Our bigger brains allow us to establish laws and cultural norms which govern the way we interact. We make up stories and agree to believe in them collectively, regardless of whether there is any evidence for them. These stories – about abstract concepts like gods and kingship, nations and ideologies, money and art - give us powerful reasons to cooperate and work together, even to die together.

Non-human primates spend hours every day grooming the other members of their tribe to reassure them that they will not sink their teeth and nails into them. It works, but it is inefficient, and means they cannot readily add new members to their tribe. Humans, by contrast, can walk past complete strangers on a crowded street without a second thought. Our superpowers are communication, and our capacity to sustain mutual belief in things which we either know to be illusory, or for which we have no evidence (like religion, monarchy, currency, democracy, and nationality). It is thanks to these abilities that we control the fate of this planet and every species on it.

(This means that the old cliché that our dominance is based on our capacity for rational thought is – unlike most clichés – untrue.)

So although it wasn't predicted in advance, in hindsight it is entirely logical that our most powerful technology, artificial intelligence, would first become available to most of us in the form of a communication device.

The way the economic singularity unfolds will probably be like that. Our attempts to forecast the impact of technological unemployment – assuming it arrives – will probably look absurd in hindsight. But when we get there, the outcome will seem not only natural, but perhaps even inevitable.

Un-forecasting

The description of a possible future that follows is not a prediction. The only thing we can be confident of is that the future will not be like this.

Instead, this timeline is intended to serve two functions. First, it is a rhetorical device to make some of the seemingly outlandish ideas in this book more palatable. The arguments that machines will automate our jobs away and that we could end up with an economy of abundance may well seem theoretical, abstract, and even implausible. Hopefully, the timeline will help make the possible future of an economic singularity seem less academic, less theoretical, and more real. And more hopeful.

Secondly, drawing up timelines like this one should help us to construct a valuable body of scenarios. Even when we know the future is unpredictable, it is still essential to make plans. There is good sense in the old cliché that failing to plan is planning to fail. If you have a plan, you probably won't achieve all of it. If you have no plan, you won't achieve anything.

In a complex environment, scenario development is a valuable part of the planning process. None of the scenarios will come true in their entirety, and many will be completely off the mark. But parts of some of them may approximate parts of the outcome. Thinking through how we would respond to a sufficient number of carefully thought-out scenarios should help us to react more quickly when we see the beginnings of what we believe to be a dangerous trend.

Super un-forecasting

The art of constructing a useful scenario is similar to forecasting, which has been extensively studied by Canadian political scientist Philip Tetlock, co-author of the book "Superforecasting: the art and science of prediction." He has found that the best forecasters share a number of traits. First, they treat their views about what will happen as hypotheses, not firm beliefs. If the evidence changes, they change their hypothesis.

Secondly, they look for numerical data. Now we all know that there are lies, damned lies and statistics, and that data is often used in public debate in the same way that a drunk uses a lamp-post: more for support than for illumination. But used carefully and honestly, data is our friend. It is after all the root of the scientific revolution that has lifted most of our species out of poverty and squalor.

Thirdly, they look for context. He cites the example of guests at a wedding, admiring the beauty and grace of the bride and the dashing good looks of the bridegroom, and assuring each other that they will share a long and happy life together. The super-forecaster is a contrarian, noting that around half of all marriages fail, and that the failure rate increases with second and third marriages, especially when one or other partner has a history of infidelity, as with the happy couple today. If she is a tactful super-forecaster, she keeps these thoughts to herself.

Ironically, super-forecasters are often not the people who get listened to in discussions about the future. We tend to pay more attention to those who speak most confidently, and offer clarity and certainty. People who equivocate and offer measured suggestions often don't cut through the noise.

So here goes, with the equivocation minimised.

2030 - Panic averted

Vehicles without human drivers are becoming a common sight in cities and towns all over the world. Professional drivers are starting to be laid off, and it is clear to everyone that most of them will be redundant within a few years. At the same time, employment in call centres and the retail industry is hollowing out as increasingly sophisticated digital assistants are able to handle customer enquiries, and the move to online shopping accelerates. The picking function in warehouses has been cost-effectively automated, and we are starting to see factories which normally have no lights on because no humans are working there.

Many companies have laid off some workers, but most have reduced their headcount primarily by natural wastage: not replacing people who retired or moved on. As a result, there have been fewer headlines about massive redundancies than some people feared, but at the same time it has become much harder for people to find new jobs. The unemployment rate among new graduates is at historically high levels in many countries.

But instead of panicking, the populations of most countries are reassured by their political and business leaders telling them that they have a plan. This is possible because a consensus formed during the second half of the 2020s that an economy of abundance is achievable, and numerous policy documents have been published explaining how it could be reached.

1. Transportation. The full automation of driving took longer than the most bullish pioneers expected, but the capability was in place by the mid-2020s. Regulation and popular resistance was less of a hurdle than some had feared, and as soon as the technology was ready, it started to be implemented in almost every developed country, and very soon afterwards in most of the developing world.

The biggest impact is being felt in the commercial vehicle sector. Fleet managers are painfully aware that their competitors are cutting out the cost of their human drivers - or preparing to - and they know that they must do likewise or go out of business. The non-driving tasks carried out by drivers, such as checking their loads, protecting them from robbery, helping to load and unload have turned out to be achievable through other means. There are still quite a few human-driven commercial vehicles, but it is clear their days are numbered.

Elon Musk was right: the early buyers of self-driving cars are offsetting the purchase cost by renting them out through apps like Uber to other people when they are not using them. But many people, especially in cities, have decided not to upgrade to a self-driving car, and rely instead on the fleets of automated taxis which are offered by partnerships between auto manufacturers and technology companies. This pooling of transportation facilitates the shift to electric cars.

2. Manufacturing. Industrial robots are cheap enough, and easy enough to programme and maintain, that manufacturers often choose to buy a robot rather than hire a human when they expand a line. The most sophisticated manufacturers of cars and electrical equipment now have a few "lights-out" factories where no humans are normally required, but this is still rare. Likewise, although a few manufacturers have laid off large numbers of workers, most have yet to make this step, relying instead on natural wastage to reduce costs.

3. Agriculture. Farmers are buying more and more robots for both crops and animal husbandry. On a growing number of farms with high-value crops, small

wheeled devices patrol rows of vegetables, interrogating plants which don't appear to be healthy specimens, and eliminating weeds with herbicides. Cattle are entirely content to be milked by robots, so fewer members of the declining population of farm workers still have to get up before daybreak every day.

4. Retail. The shift towards purchasing goods and services online continues, and there is growing automation within shops. In many supermarkets, shoppers no longer have to unload and re-load their trolleys: the goods are scanned when removed from the shelf, or while inside their baskets. Fewer attendants are required in the checkout area. In fast food outlets, so-called "McJobs" are disappearing as burgers and sandwiches are assembled and presented to customers without being handled by a human.

5. Construction. Building firms are using more pre-fabricated units, but the conditions on construction sites remain highly variable and unpredictable. Robots which can handle this unpredictability are still too expensive to replace humans. There are experiments with exoskeletons for construction workers, but these are also expensive.

6. Technology. The tech giants are still fighting to recruit and retain machine learning experts; the salaries and bonuses offered were previously unknown outside financial services and professional sports. There has been lively debate about whether two of the cloud-

based platforms should be broken up on monopoly grounds, but calls to do so have so far been denied because there was no evidence that the platforms were actually operating to the detriment of the consumer. The platforms were also able to establish that being broken up would destroy the linkages that made them so powerful, and so beneficial to consumers.

7. Utilities. Water companies and power generation and transmission firms are building out fleets of tiny robots and drones which patrol pipes and transmission lines, looking for early warning signs of failure.

8. Finance. Retail banking is mostly automated and web-based, and consumer feedback on the quality of service is improving. Wealthy people now get some of their investment advice directly from automated systems, but human investment advisers still serve most of the market. In corporate finance, human advisers show no signs of being replaced, although their back office systems are heavily automated.

9. Call centres. Enquiry handling that was offshored to India and the Philippines, and then repatriated to home countries, is now being offshored again - this time to machines housed in cold climates where the cost of keeping the servers cool are lower.

10. Media and the arts. Virtual and augmented reality equipment is now very impressive, but still improving fast. Spending time in VR arcades is one of the

most popular pursuits for teenagers and young adults. But despite numerous false dawns, AR and immersive VR for consumers is still not good enough for the mass market, and smart glasses remain a niche business for enterprise applications. As usual, porn and sport look like being the killer apps for consumer VR, but there are unexpected hits too, such as light-hearted "how-to" shows about parenting and relationship enhancement.

11. Management. Managers whose job consisted of processing information and passing it on are looking for (and struggling to find) new work, but many managers who deal with other people – staff or customers – are still employed.

12. Professions. The tedious jobs which traditionally provided training wheels for accountants and lawyers ("ticking and bashing" for auditors and "disclosure" or "discovery" for lawyers) are increasingly being handled by machines. Skeptics about technological unemployment point out that the amount of work carried out by professional firms has actually increased, as whole categories of previously uneconomic jobs have become possible, and that professionals are kept busy because the machines still need training on each new data set. But fewer trainees are being hired, and thoughtful practitioners are writing articles in their trade magazines asking where tomorrow's qualified lawyers and accountants will come from.

13. Medical. AIs are ingesting data sent by patients

from their smartphones and carrying out triage. Sometimes they respond with simple diagnoses and treatment recommendations, sometimes they pass the enquiry to a human doctor. Medical professionals and regulators are nervous about these experiments, but there is growing evidence of positive outcomes. Overall, the workload for the medical profession has increased rather than reduced, as patients are taking a more active and better informed approach to their own healthcare.

Hospitals in Japan are using robot nurses to great effect, and these trials are followed with great interest elsewhere. Pharmaceuticals designed to raise the IQ of adults are in clinical trials.

14. Education. There is now overwhelming evidence that techniques such as flipped learning and competency-based learning produce impressive outcomes, but they are still far from being universally adopted. Competitive environments, such as the UK's private school system, are experimenting with AI-assisted learning, in which students have personal AI tutors. They are rudimentary at present.

15. Government. There is a worldwide drive to get most government services delivered online, and cheaper.

2040 - Transition
Large numbers of people are now unemployed, and

welfare systems are much enlarged almost everywhere. Pretty much everyone now accepts that Universal Basic Income is not the solution, because there is no point paying generous welfare to the many people who remained in lucrative employment, and because it is widely agreed that a basic income is insufficient for the rest. The new welfare has many incarnations, and interesting new experiments are springing up all the time. Concepts like PCI (Progressive Comfortable Income), and HELP (Human Elective Leisure Programme) are being discussed not only in the think tanks, but in kitchens, bars and restaurants all over the world.

1. Transportation. Most jurisdictions now have roads which are off-limits to human drivers. Insurance premiums have plummeted, and fears about self-driving cars being routinely hacked have not been realised. A vocal minority of citizens (which, to general surprise, comprises equal numbers of men and women) are scathing about this arrangement, dubbing the communal cars THEMs, ("tedious horizontal elevator machines").

Professional drivers are now extremely rare. Their disappearance was resisted for a while – sometimes fiercely. Autonomous vehicles were frequently attacked in some places, a favourite tactic being to spray-paint the cameras and LIDARs they rely on. Some high-profile arrests and jail sentences quickly put a stop to the practice.

Deliveries of fast food and small parcels in major cities are now mostly carried out by autonomous drones, operating within their own designated level of airspace. Sometimes the last mile of a delivery is carried out by autonomous wheeled containers. For a while, teenagers delighted in "bot-tipping", but with all the cameras and other sensory equipment protecting the bots, the risk of detection and punishment became too high.

Many years after they first became an emblem of the future, flying cars finally became a real phenomenon. They fly through designated air corridors between major cities, generally taking off from and landing on the roofs of tall buildings. Propelled by multiple electric-powered rotors, they are mercifully quiet.

2. Manufacturing. Many large factories and warehouses are now dark: no light is required because no humans work there. People are becoming a rarity in smaller sites too.

3D printing has advanced less quickly than many expected, as it remained more expensive than mass production. But it is common in niche applications, like urgently required parts, components with complex designs, and situations where products are bespoke, as in parts of the construction industry. They have an impact on businesses and the economy far greater than their modest output level would suggest.

3. Agriculture. Farmers are moving heavily into leisure services, as their families and staff are losing their roles to robots.

4. Retail. Online shopping reaches 75% of all retail purchases, with a small but growing number of items being 3D printed domestically or in neighbourhood facilities, often with an element of customisation by the consumer. In the High Street shops that remain, human shop assistants are being replaced by robots, except in high-margin sectors where humans help to create an experience rather than merely facilitating transactions.

5. Construction. Human supervision is still the norm for laying foundations, but pre-fabricated (often 3D-printed) walls, roofs and whole building units are common. Robots, and humans in exoskeletons are increasingly used to assemble them. Drones populate the air above construction sites, tracking progress and enabling real-time adjustments to plans and activities.

6. Technology. Smart glasses have replaced smartphones, and "wearables" are commonplace - electronic devices that are worn on the body as accessories, or woven into clothing. The first "inside-ables" are appearing, and have been made fashionable by Lord Beckham, the ex-footballer and fashion magnate. The Internet of Things is well established, with everyone receiving messages continuously from thousands of sensors and devices implanted in vehicles, roads, trees, buildings, etc. Fortunately, the messages are intermediated by personal digital assistants, mostly instantiated in smart glasses. These digital assistants have acquired the generic name of "Friends", but their owners often endow them with pet names.

New types of relationship and etiquette are evolving to govern how people interact with their own and other peoples' Friends, and what personalities the Friends should present. Competition between brands to provide the best Friends software is fierce.

There is lively debate about the best ways to communicate with Friends and other computers. Concealed microphones handle much of the traffic, but millions of people are also learning how to use small, one-handed keyboards which liberates them from traditional keyboards at times when voice is inappropriate. Some believe these new keyboards will be quickly superseded by Brain-Computer Interfaces (BCI), but this has made less progress than its early enthusiasts expected.

Another promising technology is tattoos worn on the face and around the throat which have micro-sensors to detect and interpret the tiny movements when people sub-vocalise, i.e., speak without actually making a noise. The tattoos are usually invisible, but some people have visible ones, which give them a cyborg appearance.

A growing amount of entertainment and personal interaction is mediated through virtual reality. Good immersive VR equipment is now found in most homes, and it is increasingly rare to see an adolescent in public outside school hours.

Polls suggest that most people now think that the arrival of artificial general intelligence (AGI) and then superintelligence is a serious possibility within a generation or two. Significant expenditure is flowing into research on how to make sure the outcome is positive, and the moral and religious implications are hotly debated.

7. Utilities. In many organisations, most operations are now automated. The main role of humans in these organisations is testing security arrangements. Several hundred people died in two significant hacking incidents – one in the US and one in Europe. This has prompted huge investment in upgraded security arrangements. In another high-profile incident, an AI system managed the disaster containment and recovery process flawlessly, and much faster than humans could have.

8. Finance. Retail banking is now fully automated, and investment advice is going the same way. Corporate financiers are in retreat, and their previously stratospheric incomes have fallen sharply.

9. Call centres. Few humans now work in call centres.

10. Media and the arts. All major movies made by Hollywood and Bollywood are now produced in VR, along with all major video games. To general surprise, levels of literacy – and indeed book sales – have not fallen. In a number of genre categories, especially romance and crime, the most popular books are written by AIs.

Long-distance communication is massively improved by VR.

Major sporting competitions have three strands: robots, augmented humans, and un-augmented humans. Audiences for the latter category are dwindling.

Dating sites have become surprisingly effective. They analyse videos of their users, and allocate them to "types" in order to match them better. They also require their members to provide clothing samples from which they extract data about their smells and their pheromones. The discovery that relationship outcomes can be predicted with surprising accuracy with these kinds of data has slashed divorce rates.

11. Management. The ranks of middle management are thinning out. Shareholders are investing heavily in Distributed Autonomous Corporations (DACs), firms consisting of unsupervised AIs which create new business models and strategies and transact with other firms without any humans in the loop.

12. Professions. Partners in law firms and accountancy firms are working shorter hours. Human intakes to these firms are dwindling. Most criminal law cases now relate to digital crime: with tiny, powerful cameras everywhere accompanied by highly effective facial recognition technology, the perpetrators of physical crime are generally caught.

13. Medical. Opposition to the smartphone medical revolution has subsided in most countries, and most people obtain diagnoses and routine health check-ups

from their "Friends" several times a week. Automated nurses are becoming increasingly popular, especially in elder care.

Several powerful genetic manipulation technologies are now proved beyond reasonable doubt to be effective, but backed by public unease, regulators continue to hold up their deployment. Cognitive enhancement pharmaceuticals are available in some countries under highly regulated circumstances, but are proving less effective than expected. There are persistent rumours that they are deliberately being engineered that way.

Ageing is coming to be seen as an enemy which can be defeated.

14. Education. Data on learning outcomes is steam-rollering resistance to new approaches. Customised learning plans based on continuous data analysis are becoming the norm. Teachers are becoming coaches and mentors rather than instructors. More and more schools are experimenting with classroom AIs.

15. Government. There is growing pressure to reduce the numbers of politicians and civil servants, as more and more government services are automated. Many jurisdictions are debating the merits of using technology to enable direct democracy, which is being pioneered by Switzerland. Most people are skeptical, fearing the tyranny of the temporary majority. Policemen in most countries record all interactions with members of the public, and public satisfaction levels with them are generally rising.

16. Charities. Non-profit organisations are enjoying a surge, thanks to an influx of talent as capable people can't find work elsewhere.

2050 - The economy of abundance

Artificial intelligence systems have driven the cost of production of most non-luxury goods and services close to zero. Few people pay more than a token amount for entertainment or information services, which means that education and world-class healthcare are also much improved in quality and universally available. The cost of energy is dramatically reduced also, as solar power can now be harvested, stored and transmitted almost for free. Transportation involves almost no human labour, so with energy costs so low, people can travel pretty much wherever they want, whenever they want. The impressive environments available in virtual reality do a great deal to offset the demand for travel that this might otherwise have created.

Food production is almost entirely automated, and the use of land for agriculture is astonishingly efficient. Vertical farms play an important role in high-density areas, and wastage is hugely reduced. The quality of the housing stock, appliances and furniture is being continuously upgraded. A good standard of accommodation is guaranteed to all citizens in most developed countries, although of course there are always complaints about the time it takes to arrive. Personalised or more luxurious versions are available at very reasonable prices to those still earning extra income. Almost no-one in developed countries now lives in cramped, damp, squalid or noisy conditions. Elsewhere in the world, conditions are catching up fast.

Other physical goods like clothes, jewellery and other personal accessories, equipment for hobbies and sports, and a bewildering array of electronic equipment are all available at astonishingly low cost. But access to most goods and services is still rationed by price. Almost nobody wants for the necessities of civilised life, but at the same time, nobody can afford to buy absolutely everything they might wish for. It is generally accepted that this is actually a good thing, as it reduces wasteful production, and it means the market remains the mechanism for determining what goods are produced, and when.

Unemployment has passed 75% in most developed countries. Among those still working, nobody hates their job: people only do work that they enjoy. Everyone else receives an income from the state, and there is no stigma attached to being unemployed, or partially employed. In most countries the citizens' income is funded by taxes levied on the minority of wealthy people who own most of the productive capital in the economy, and in particular on those who own the AI infrastructure. The income is sufficient to afford a very comfortable standard of living, with access to almost all digital goods being free, and most physical goods being extremely inexpensive.

In many countries, some of the wealthy people have agreed to transfer the productive assets into communal ownership, controlled by decentralised networks using blockchain technology. Those who do this enjoy the sort of popularity previously reserved for film and sports stars.

Some countries mandated these transfers early on by effectively nationalising the assets within their legislative reach, but most retreated from this approach when they realised that their economies were stagnating, as many of their most innovative and energetic people emigrated. Worldwide, the idea is gaining ground that private ownership of key productive assets is distasteful. Most people do not see it as morally wrong, and don't want it to be made illegal, but it is often likened to smoking in the presence of non-smokers.

The gap in income and wealth between rich and poor countries has closed dramatically. This did not happen because of a transfer of assets from the West to the rest, but thanks to the adoption of effective economic policies, the eradication of corruption, and the benign impact of technology in the poorer countries.

Another concern which has been allayed is that life without work would deprive the majority of people of a sense of meaning in their lives. Just as amateur artists were always happy to paint despite knowing that they could never equal the output of a Vermeer, so people now are happy to play sport, write books, give lectures and design buildings even though they know that an AI could do any of those things better than them.

Not everyone is at ease in this brave new world, however. Around 10% of the population in most countries suffers from a profound sense of frustration and loss, and either succumbs to drugs or indulges almost permanently in escapist VR entertainment. A wide range of experiments is under way around the world, finding ways to help these people join their friends and families in less destructive or limiting lifestyles. Huge numbers of people outside that 10% have occasional recourse to therapy services when they feel their lives becoming slightly aimless.

Governments and voters in a few countries resisted the economic singularity, seeing it as a de-humanising surrender to machine rule. Although they found economically viable alternatives at first, their citizens' standard of living quickly fell far behind. Several of these governments have now collapsed like the communist regimes of Eastern Europe in the early 1990s, and the rumours that President-for-life Putin met a very grisly end in 2041 have never been substantiated. The other hold-outs look set to follow – hopefully without violence.

1. Transportation. Humans do not drive vehicles on public roads, and very few commercial vehicles have human attendants. Young people do not take driving tests. Humans do drive vehicles for sport, but even motor sports are now mostly competitions between self-driving cars. Many people now live nomadic lifestyles in RVs. They spend a few days in one place and then, often overnight, their automated RVs drive to them to their next destination. The choice of destination is often governed by who else is going to be there, or a special event that is of interest. Selecting the next

destination and arranging who to meet there is usually mediated by AIs, as is the route planning and driving.

2. Manufacturing. Almost all factories and warehouses are dark. 3D printing still accounts for a minority of the total goods manufactured, but it is becoming competitive with some forms of mass production.

3. Agriculture. Robots do most farm work.
Some countries have large communally-owned agricultural processing concerns which send out meals on drones in a service often described as Netflix for food.

4. Retail. The great majority of items are now bought online. Retail outlets on High Streets and city centres are mostly experiential rather than transactional, and mostly staffed by AIs and robots.

5. Construction. Robots now carry out most of the work on construction sites.

6. Technology. Since AI provides a large proportion of the value in most products and services, there is a major concentration of capital and wealth in the hands of shareholders and key employees in this sector. Its foremost talent is now applied to developing artificial general intelligence and making sure that it is safe for humans. The Internet of Things is all-pervasive, and the environment appears intelligent.

The companies that provide "Friends" have been

obliged to make them open-source. Friends are so critical to everyone's lives that being restricted to any one company's walled garden was unacceptable.

7. Utilities. Overwhelmingly automated.

8. Finance. Overwhelmingly automated.

9. Call centres. Unchanged.

10. Media and the arts. In sports, robot competitions now generate larger audiences than their human counterparts. The International Olympics Committee de-lists the human versions of around half of all sports.

Haptic body suits combined with VR headsets now provide truly immersive virtual environments. Counselling (by AIs) is required by a section of the population who struggle to maintain the distinction in their minds between reality and VR.

To general surprise, people still read books, but they are very different products now, with holographic illustrations, and often with several alternative story lines developed by their AI authors, which readers can choose between.

Dating sites are now mostly accessed by personal digital assistants. "My Friend likes your Friend" was a standard conversation opener for a while.

11. Management. Many companies now consist of just a few strategists, whose main role is to forecast the optimal business model for the next financial quarter,

but they are struggling to keep up with their AI advisers.

12. Professions. Accountancy and the law are largely automated.

13. Medical. Demand for human doctors is dwindling and professional nursing has been almost entirely automated. Everyone in developed economies has their health monitored continuously by their "Friends". Most people spend a certain amount of time each week visiting family, friends and neighbours who are unwell, just to converse. Sick and disabled people are greatly comforted by their relationships with talking AI companions, some resembling humans, others resembling animals.

Significant funds are now allocated to radical age extension research, and there is talk of "longevity escape velocity" being within reach – the point when each year, science adds a year to your life expectancy. Most forms of disability are now offset by implants and exoskeletons, and cognitive enhancements through pharmaceuticals and brain-computer interface techniques are showing considerable promise.

14. Education. The sector has ballooned, with many people now regarding it as recreation rather than work. Most education is provided by AIs.

15. Government. Safeguards have now been found to enable direct democracy to be implemented in many

areas. Professional politicians are now rare.

CHAPTER 30

SUMMARY AND CONCLU-
SIONS

Automation and unemployment

We cannot be certain, but it looks likely that improvements in machine intelligence over the next few decades are going to make it impossible for most humans to earn a living. We would be wise to devote some resource to working out how to deal with this development in case it does happen – indeed we would be foolish not to.

During the industrial revolution, concerns that automation would lead to permanent mass unemployment turned out to be unfounded – unless you were a horse. (The Engels pause was lengthy, but not permanent.) Instead, automation raised productivity and output across the economy. The unfounded concerns became known as the Luddite fallacy.

In the information revolution, mankind's fourth

great wave of transformation (although definitely not the fourth industrial revolution), machines are increasingly able to out-perform humans in cognitive tasks. This is likely to put humans in the predicament that horses were placed in by the industrial revolution.

In 1900, 40% of American workers were employed in agriculture, and that has now fallen below 3%. The farm workers found better jobs elsewhere in the economy, sometimes in occupations which their parents could not have imagined. But horses didn't. 1915 was "peak horse", with 21.5 million horses in America pulling vehicles on roads, and machinery on farms. The horse population in America today is two million. The difference between the horses and the humans is that when machines took over the muscle jobs, humans had something else to offer: our cognitive, emotional and social abilities. Horses had nothing else to offer, and their population collapsed.

Past rounds of automation have mostly been mechanisation, the replacement of human and animal muscle power. The coming waves of automation will substitute our cognitive abilities. Machines don't need to become human-level to displace most of us from our jobs. They simply have to become better than us at what we do for a living. They are close to or at parity with us with regard to many forms of pattern recognition, including image recognition and speech recognition. They will soon be considerably better than us, and continuing to improve at an exponential rate. They are catching up on our ability to process natural language, and they will overtake us there as well.

Once a machine can do your job, it will quickly be able to do it faster, better and cheaper than you can. Machines don't eat, sleep, get drunk, tired or cranky. And unlike human brains, their abilities will continue to improve at an exponential rate.

Skeptics

Skeptics offer two arguments to prove that automation cannot cause lasting widespread unemployment. Neither is persuasive.

First, they claim that since automation has not caused lasting mass unemployment in the past, it cannot do so in the future. But past performance is no guarantee of future outcome. If it was, we would never have learned how to fly. The observation that past rounds of automation have not caused lasting mass unemployment is not even true – if you consider the horse.

Secondly, some skeptics claim that since human wants and needs will never be fully satisfied, there will always be jobs for humans. This is also a weak argument: our needs may never be fully satisfied, but that does not mean humans will be able to compete with the machines in the attempt to do so.

In addition to these two arguments, skeptics offer some explanations of the kind of work we will all be doing when machines have taken over many or most of our existing jobs.

First, they suggest that we will race with the machines instead of against them, becoming "centaurs", and occupying ourselves with the "icebergs" of new work which machines have made possible. But these are only likely to be temporary respites, as the machines continue their rapid improvement.

Then they suggest that we will all do the caring jobs which machines cannot do because they have no consciousness and therefore no empathy. This turns out to be unconvincing too: people actually like being looked after by machines. They also suggest that we will buy goods and services from humans rather than machines out of chauvinism, and that we will all be artisans, or entrepreneurs. None of these ideas stand up to scrutiny.

Finally they invoke the magic job drawer, out of which will fly all sorts of jobs which we cannot imagine today because the technologies to make them possible have not been invented yet. They claim this is what happened during the industrial revolution: an agricultural labourer in 1800 could not have imagined that his descendant would become a web experience designer. But the idea that we all do jobs today that our grandparents could not have imagined is a myth. 80% of the jobs people do today are jobs which existed in 1990. And even if we do invent all sorts of new jobs which are unimaginable today, whatever new jobs we do invent, the machines will probably take over most of them as well. In the medium term.

In short, it is highly likely that within a generation or so, a large minority of people – perhaps the majority – will not be able to earn a living through jobs.

It is true that many well-informed people are skeptical of the technological unemployment thesis. They include economists like David Autor, David Dorn, and Robert Gordon, as well as technology industry leaders like Eric Schmidt, Sundar Pichai and Marc Andreessen. But arguments from authority should always be viewed with suspicion, and there are plenty of equally well-informed people on the other side of the debate, including AI researchers like Stuart Russell and Andrew Ng, and technology industry leaders like Mark Zuckerberg, Elon Musk and Sam Altman.

The upside

Fortunately, technological unemployment, if it happens, does not have to be bad news; in fact it should be extremely good news. Some people are lucky enough to love their jobs, and find fulfilment in them. For many more people, work is simply a way to generate an income for themselves and their families. It may provide a purpose, but it does not provide meaning. A world in which machines do all the boring work could be wonderful. They could be so efficient that goods and services could be plentiful, and in many cases free. Humans could get on with the important business of playing, relaxing, socialising, learning and exploring. Surely this is what we should be aiming for.

People who believe that technological unemployment can lead to this world of freedom are the optimists. People who believe that humans will have to remain in paid employment forever are the true pessimists.

Challenges

Reaching this positive outcome requires us to confront a number of challenges, including meaning, economic contraction, income, and allocation. Although we probably have a considerable amount of time before these challenges bite hard in practice, the possibility of a panic erupting when the majority of people start to anticipate the likely changes means that the problem is more urgent than it might at first appear.

When people first take seriously the idea of the economic singularity, they often fear that unemployed humans will find their new lives hollow, lacking in meaning, and perhaps even boring. This fear is probably mostly misplaced. For centuries, aristocrats in most countries didn't work for a living, and in many societies they viewed work as a demeaning activity, to be avoided by "people of quality". Some of them got into trouble with drink, drugs and gambling, but only a small minority. Most of them seem to have led contented lives, however questionable we might find the economic systems they operated in.

Likewise, retirement is rarely considered a disaster in developed countries – as long as you retire with a sufficient income. Even though most of us only get to enjoy it when we are past our prime, most retirees find enough projects and pastimes to keep themselves busy and at peace. Numerous surveys have found that happiness is U-shaped: we are at our most content during childhood and retirement, and it is probably no coincidence that these are the periods in our lives when we don't work for a living. If we retired when still in our prime, we would be even better equipped to enjoy our lives of leisure.

The first really big challenge posed by the economic singularity is how to ensure that everyone has a good income - or rather, full access to the goods and services that are needed for a good life. (This overshadows and subsumes the problem of economic contraction.)

Many people think the answer is universal basic income, but in an economy where goods and services remain expensive, UBI will either be set too low to be sufficient, or it will be unaffordable. It cannot be paid for by soaking the rich, by eradicating bureaucracy, or by taxing robots. Two of its major weaknesses are clearly stated in its very name: a universal payment wastes enormous sums on people who don't need it, and a basic income is inadequate, and would not create a stable, sustainable society.

The effective way to solve the income problem is to drive close to zero the costs of all the goods and services required for a good life. In other words, to achieve an economy of abundance. Although this seems implausible at first sight, there are signs that it is in fact achievable.

If we do solve the income problem, there will still remain a problem of allocation: in a world where most people cannot vary their income, how do we decide who enjoys the scarce things in life, like the desirable detached houses in the better neighbourhoods, the penthouse apartments, the beach-front properties, the original Vermeers and the Aston Martins? Virtual reality might provide at least some of the answer.

The last challenge we considered was the possibility of speciation. Inequality is often over estimated as a contemporary social evil, but the continuing acceleration of technological development might change that. Enhancements to our physical and cognitive performance will come along faster and faster, and groups with privileged access to them may start to diverge from everyone else, and become a separate species. The author Yuval Harari has referred to this scenario in the chilling phrase, "the gods and the useless". The "Brave New World" depicted by Aldous Huxley in 1931 might be one of the least bad outcomes of this scenario.

If the post-economic singularity world needs a different type of economy, then we need to start thinking now about what that might be – and also how to get there. The damage that could be caused by an uneven or violent transition to the new world could be immense.

Scenarios

In chapter 28, we considered four potential outcomes to this process: no change, full employment, social collapse, and the Star Trek economy.

Robert Gordon and John Markoff claim to believe that the great age of innovation is behind us, and technological progress today is just twiddling with unimportant apps. Estimable though these writers are, this is obvious nonsense. Google Search plus Wikipedia has given us something which our ancestors would have mistaken for omniscience, and it is clear to almost everyone that the advent of smarter and smarter machines is going to have enormous impacts on all of us.

Full employment is a more plausible scenario, but the case that it will certainly be true has not been made. It is hard to see why, given their continued exponential improvement, machines will not be able to replace humans in most paid work roles: it is just a matter of time.

The scenario of social collapse, or of societies falling prey to some kind of totalitarian control, is all too plausible. But it is obviously not acceptable: we must make sure it does not happen.

Which leaves the economy of abundance. If it can be achieved – and achieved in time, it offers the potential for humans to flourish in ways that our ancestors could barely have dreamed of. It seems to be the best solution to the challenges posed by the economic singularity, and it is what we should be aiming for.

We need a plan

At a conference in June 2017,[422] Professor Stuart Russell suggested that we should lock a group of economists and science fiction authors into a room and not let them out until they have come up with a plan for how to respond to technological unemployment. It's a terrific idea. We need a plan for how to navigate the economic singularity, and devising the plan will require both intellectual rigour and free-wheeling creativity and imagination.

Scarcely a day goes by without an article appearing in the media debating whether robots will steal our jobs. Often they are impressionistic, poorly-argued pieces - it is depressing, for example, how often the Reverse Luddite Fallacy rears its ugly head. There are worryingly few places where sustained effort is being applied to solving the problem that these articles discuss. The best-known is the Oxford Martin Programme on Technology and Employment,[423] led by Dr Carl Benedikt Frey and Michael Osborne, who wrote the influential 2013 paper which suggested that 47% of American jobs would be displaced.

There are at least four permanently-established organisations studying the risk to humanity posed by the potential arrival of superintelligence.[424] This is a very good thing, and the technological singularity is an existential threat to humanity, which the economic singularity is probably not. But the economic singularity is highly likely to arrive considerably sooner, and seeking solutions to the challenges it poses should be a priority.

We need to establish research institutes and think tanks to study the issues, to draw in diverse opinions, to canvas ideas and challenge them. And we need to start doing this now.

Cognitive automation has not yet begun to cause technological unemployment – or at least, not to any significant extent. The machine learning "big bang" happened less than a decade ago, and the enormous power of deep learning has yet to be harnessed by many organisations outside the tech giants. We have time to solve this problem, but we do not have time to waste. We should be monitoring developments, and drawing up forecasts and scenarios.

The view that most people will be rendered unemployable by machine intelligence within the next few decades is probably a minority opinion at the moment, but many of the people best-placed to understand what is coming do see it as a likely future.

Of course there are many other causes whose proponents would make the same claim on the limited resources at our disposal: climate change, inequality, and ocean pollution to name just three. Even the largest and wealthiest nations cannot expend sufficient resources to de-risk every danger that has been identified by somebody somewhere as important. We have to prioritise. I hope this book has persuaded you that the economic singularity should be high on that list of priorities.

Monitoring and forecasting

One of the core tasks of these think tanks and research institutions would be to monitor developments within the major economies of the world, and alert us all when significant trends become apparent. But working empirically with data is not enough. The challenges we want to rise to are in the future, not the past: there simply is no data about them yet. We need to forecast as well as monitor, in spite of the well-known problems with forecasting.

Prediction markets are probably part of the answer. People make their best estimates when they have some skin in the forecasting game. Offering people the opportunity to bet real money on when they see their own jobs or other peoples' jobs being automated may be an effective way to improve our forecasting. It also harnesses the wisdom of the crowd.

In a prediction market, someone asks a question, often one requiring a yes / no answer, such as, "Will self-driving technology replace half of New York's taxi drivers by 2040?" Other people can buy and sell contracts for yes, or contracts for no. The yes contracts are paid if the answer to the question turns out to be yes, and the no contracts are paid if the answer is no. The price of the contracts move, determined by supply and demand, and they are a prediction of the probability of the event occurring. So if the yes contract is priced in the market at 80, that means the market thinks there is an 80% chance of half of New York's taxi drivers being replaced by 2040. The market ends in 2040, or when half the drivers are replaced, whichever happens first. If they are replaced, the price of the contract goes to 100, otherwise it goes to zero.[425]

Scenario planning

Obviously we are lacking sufficient information to draw up detailed plans for the way we would like our economies and societies to evolve. But we can and should be doing detailed scenario planning.

Scenario planning has been practised by military leaders since time immemorial. It was given the name by Herman Kahn, who wrote narratives about possible futures for the US military while working for the RAND Corporation in the 1950s. (His suggestion that a nuclear war might be both winnable and survivable made him one of the inspirations for Dr Strangelove in the classic 1964 movie.[426]) Scenario planning was adopted by Shell after it (along with the rest of the oil industry) was disastrously wrong-footed by the rise of the oil cartel OPEC in the 1970s.[427]

Scenario planning is more art than science, and it cannot prevent forecasts being mistaken, but it is a valuable discipline. When we commit our thoughts about a possible future to paper we are forced to consider them rigorously. Think tanks and research institutes doing this work could make a valuable contribution.

An informal version of this is the daily business of futurists and futurologists, people who are often viewed with skepticism by the wider public. Perhaps that will change – in fact, perhaps futurology will come to be seen as a mission-critical profession. Science fiction writers also have an important role, in providing vivid metaphors and warnings.

Global problem, global solution

Technological unemployment will affect every country, albeit at different rates and perhaps in different ways. We are in a race against time to find solutions to the problems it raises, but not in a race against each other. We can and should co-operate. At the moment, on the rare occasions when politicians talk about AI, they usually talk about making sure their country is in the lead in some kind of an AI race, or at least not falling too far behind. Certainly, it makes sense for every government to seek to promote the domestic development of AI, as that will benefit their people economically. But when it comes to the big issues - the singularities – they should be working together, sharing ideas, learning from each other.

Upsides, downsides, and timing

The technological unemployment thesis might turn out to be false, with humans remaining in employment for decades to come. If we allocate the resources to establish a few think tanks and research institutes, and the problem they were designed to tackle never arises, what have we lost? A few million dollars at the very most.

If, on the other hand, the technological unemployment thesis turns out to be correct, but we do nothing to prepare, some kind of panic would be almost inevitable, leading to extreme versions of populism, and perhaps dictatorships. The outcome would be very unlikely to be appetising.

The skeptics about the technological unemployment thesis might be right, but they have no compelling arguments, and they may very well be wrong. Just hoping that they are right does not look like a wise policy.

If the optimists are right, and a leisure society funded by an economy of abundance is possible, then the prize is well worth the modest investment required to attain it, and to make the transition smoother.

APPENDIX 1: PROPONENTS

(WRITERS WHO TAKE THE PROSPECT OF TECHNOLOGICAL UNEMPLOYMENT SERIOUSLY)

Frey and Osborne

Carl Benedikt Frey and Michael Osborne are the directors of the Oxford Martin Programme on Technology and Employment.[428] Their 2013 report "The future of employment: how susceptible are jobs to computerisation?" has been widely quoted. Its approach to analysing US job data has since been used by others to analyse job data from Europe and Japan.

The report analyses 2010 US Department of Labour data for 702 jobs, and in a curious blend of precision and vagueness, concludes that "47% of total US employment is in the high risk category, meaning that associated occupations are potentially automatable over some unspecified number of years, perhaps a decade or two." 19% of the jobs were found to be at medium risk and 33% at low risk. Studies which have extended these findings to other territories have yielded broadly similar results.

The methodology overlays rigour on guesswork. 70 of the jobs were categorised in a brainstorming session, and these categorisations were then extended to the other 632 jobs using calculations which will mystify anyone with only school-level maths, including a statistical tool called Gaussian process classifiers. But it would be unfair to criticise the report for lack of rigour. Forecasting is not an exact science; the authors adopted the most scientific approach they could devise, and made no attempt to hide its subjective elements.

As well as sounding the alarm about the possibility of technological unemployment, the report suggests that the "hollowing out" of middle class jobs will stop. A 2003 paper by David Autor (of whom, more below) observed that income has increased for high earners and (albeit less rapidly) for low earners, but stagnated for medium-level earners. Maarten Goos and Alan Manning characterised this hollowing out as the favouring of "Lovely and Lousy jobs". Frey and Osborne argue that in the future, susceptibility to automation will correlate negatively with income and educational attainment, so the Lousy jobs will also disappear. They suggest that people will have to acquire creative and social skills to stay in work, but they don't appear to think that many of us will be able to change the fate that our employment history has assigned to us.

Frey and Osborne followed the 2013 report with another in February 2015, written in collaboration with senior bankers from Citibank. It provides insights into the impact of automation in a number of industry sectors, including stock markets, where the move from trading floors to digital exchanges reduced headcount by 50%. At first glance it is surprising to see bankers suggesting increased taxation to provide income for the unemployed, but it also seems they have little faith in it happening: "Such changes in taxation would seem sensible to us, but they would also be a reversal of the trends of the last few decades." They don't hold out much more hope for their other principal suggested remedy: "education alone is unlikely to solve the problem of surging inequality, [but] it remains the most important factor."

Frey and Osborne are often regarded as proponents of the idea that technological unemployment is coming soon, which is why they are included in this appendix, and not the next one. But in fact, their 2013 report left open the question of whether enough new jobs would be created to replace the jobs lost to automation. In conversation in 2020, Frey said that in fact he does not expect technological unemployment to arrive this century.[429]

In 2019, Frey published "The Technology Trap", which argues that although the Industrial Revolution created unprecedented wealth and prosperity over the long run, the immediate consequences of mechanization were devastating for large swathes of the population. Middle-income jobs withered, wages stagnated, the labour share of income fell, profits surged, and economic inequality skyrocketed. Frey argues that these trends broadly mirror those in the current wave of automation.

Martin Ford

Martin Ford is the author of perhaps the best book published so far about the technological unemployment thesis. "The Lights in the Tunnel" (2009) provoked fierce debate, and his follow-up, "Rise of the Robots" (2015) fleshed out his arguments, and responded to the criticism which the first book attracted. Awarding it the 2015 Financial Times and McKinsey Business Book of the Year, Lionel Barber, the Financial Times' editor, called it "a tightly-written and deeply-researched addition to the public policy debate … The judges didn't agree with all of the conclusions, but were unanimous on the verdict and the impact of the book."

Ford is well-placed to talk about what technology will do to the world of work. He has a quarter-century of experience in software design, and he lives and works in Silicon Valley, where he runs a software development company. His writing is calm and measured, with an engaging humility.

Ford opens "Rise of the Robots" with a dramatic illustration of the power of exponential growth. He asks us to consider driving a car at five miles an hour, and then doubling our speed 27 times over. The resulting speed would be 671 million miles an hour – fast enough to travel to Mars in five minutes[430]. This, he points out, is the number of doublings that computer power has gone through since the invention of the integrated circuit in 1958.

The book argues that AI systems are on the verge of wholesale automation of white collar jobs – jobs involving cognitive skill such as pattern recognition and the acquisition, processing and transmission of information. In fact it argues that the process is already under way, and that the US is experiencing a jobless recovery from the Great Recession of 2008 thanks to this automation process. Ford claims that middle class jobs in the US are being hollowed out, with average incomes going into decline, and inequality increasing. He acknowledges that it is hard to disentangle the impact of automation from that of globalisation and off-shoring, but he remains convinced that AI-led automation is already harming the prospects of the majority of working Americans.

In fact, since Ford's book was published the US employment figures have improved considerably, and the unemployment rate hovers around 5%, which is considered close to full employment. However, many middle-class Americans do feel squeezed, having been obliged to accept part-time work, or having missed out on wage rises. This suggests that technological unemployment has not yet begun to really bite, but we might be seeing the early warning signs.[431]

Ford pauses to review the prospects for disruption of two sectors of the economy which have so far been relatively unscathed by the digital revolution – education and healthcare. Although there is fierce resistance to the replacement of human activity by AIs in these areas – for instance in essay marking – Ford argues that no industry can ignore for long the benefits of cheaper, faster, more reliable ways of providing their products and services. He goes on to point out that the companies and industries which today are nascent and fast-growing, and tomorrow will be economic giants, are extremely parsimonious employers of humans.

The final chapters of "Rise of the Robots" explore the consequences of the trends which Ford has described. Can an economy thrive and grow if a large minority of people cannot find sufficient work to give themselves and their families a decent life? Would the consequent rise in inequality be economically harmful? More fundamentally, how will these unemployed or under-employed people make ends meet? To Ford, the answer to this last question is clear: governments will need to raise the taxes paid by those who are still working to provide an income for those who are not. But he is acutely aware of the political difficulties that this proposal faces: "American politicians are terrified to even utter the word 'tax' unless it is followed immediately by the word 'cut'."[432]

In fact, Ford seems daunted by the situation: "The political environment in the United States has become so toxic and divisive that agreement on even the most conventional economic policies seems virtually impossible," he writes. "A guaranteed income is likely to be disparaged as 'socialism'", and "The decades-long struggle to adopt universal health coverage in the United States probably offers a pretty good preview of the staggering challenge we will face in attempting to bring about any whole-scale economic reform."

Ford thinks that most people will probably still be able to find some form of paid employment – just not enough to make a decent living. Unwilling to give up on traditional American ideals like the free market, a capitalist economy and indeed the Protestant work ethic, he advocates a universal basic income of only $10,000 a year - a level low enough to leave the incentive to find work in place. Even so, he is pessimistic about the prospect of persuading his fellow Americans to adopt the idea: "a guaranteed income will probably remain unfeasible for the foreseeable future."

Erik Brynjolfsson and Andrew McAfee

As a pair of MIT professors[433], Brynjolfsson and McAfee bring academic credibility to their book on AI automation, "The Second Machine Age". They have helped to validate the discussion of the possibility of technological unemployment.

Their book (and their argument) is in three parts. The first part (chapters 1 to 6 inclusive) describes the characteristics of what they call the second machine age. They warn readers that their recitation of recent and forthcoming developments may seem like science fiction, and their prose is sometimes slightly breathless: even tenured professors can get excited about the speed of technological change and the wonders it produces.

The second part of the book (chapters 7 to 11) explores the impact of these changes, and in particular two phenomena, which they label "bounty" and "spread". "Bounty" is the "increase in volume, variety and quality, and the decrease in cost of the many offerings brought on by technological progress. It's the best economic news in the world today." This part of the book could have been written by Peter Diamandis, author of "Abundance" and "Bold".

"Spread" seems to be a synonym for inequality, although the authors are strangely reluctant to use that word.[434] It is "ever-bigger differences among people in economic success". This part of the book could have been written by a member of the Occupy movement[435]. "Spread is a troubling development for many reasons, and one that will accelerate in the second machine age unless we intervene."

Brynjolfsson and McAfee pose the question whether bounty will overcome the spread. In other words, will we create an economy of radical abundance, in which inequality is relatively unimportant because even though a minority is extraordinarily wealthy, everyone else is comfortably off? Their answer is that current evidence suggests not. Like Martin Ford, they think the American middle class is going backwards financially, and they think this trend will continue unless remedial action is taken.

So the third and final part of the book discusses the interventions which could maximise the bounty while minimising the spread. In particular, Brynjolfsson and McAfee want to answer a question they are often asked: "I have children in school. How should I be helping them prepare for the future?"[436] They are optimistic, believing that for many years to come, humans will be better than machines at generating new ideas, thinking outside the box (which they call "large-frame pattern recognition") and complex forms of communication. They believe that humans' superior capabilities in these areas will enable most of us to keep earning a living, although they think the education system needs to be re-vamped to emphasise those skills, and downplay what they see as today's over-emphasis on rote learning. They praise the Montessori School approach of "self-directed learning, hands-on engagement with a wide variety of materials ... and a largely unstructured school day." They also have high hopes for digital and distance learning, which use "digitisation and analytics to offer a host of improvements."[437]

Brynjolfsson and McAfee offer a series of further recommendations which they say are supported by economists from across the political spectrum: pay teachers more, encourage entrepreneurs, enhance recruitment services, invest in scientific research and infrastructure improvements, encourage immigration by the world's talented migrants, and make the tax system more intelligent.

These seem somewhat unremarkable proposals, although the immigration component could raise hackles in these populist times. The authors acknowledge that their effectiveness may peter out as the 2020s progress, and machines become even smarter. Looking further ahead, they warn against any temptation to try to arrest the progress in AI, and also against any temptation to move away from the tried and tested economic system of capitalism, which they claim (paraphrasing Churchill's quip about democracy again) "is the worst form of [economy] except for all the others that have been tried."[438]

The authors are very keen on Voltaire's dictum that "work saves a man from three great evils: boredom, vice and need." They are therefore wary of universal basic income, believing that an absence of work will engender boredom and depression. Instead, they argue for a negative income tax, which incentivises work. With a negative income tax of 50%, if you earn a dollar, the government gives you an additional 50 cents. They cast around for ways to keep us all in work, and rather tentatively suggest a range of exotic schemes, such as a cultural movement to prefer goods made by humans rather than machines.

Like Frey and Osborne, Brynjolfsson and McAfee leave the reader uncertain about whether they expect full-blooded technological unemployment. In conversation in 2019, Brynjolfsson told me that he thought it was possible, but not for quite a few decades. He thought it more important to focus on how to help societies thrive during the prior period of rapid job change (what I call the "Churn") rather than what may happen next.

In 2017 Brynjolfsson and McAfee published a new book called "Machine, Platform, Crowd". Like its predecessor it is written in an engaging style, and it is an entertaining and illuminating tour of some of the big changes happening in the economy. Although the first third is devoted to AI and related technologies, this time the authors steer even more firmly away from the possibility of technological unemployment. Although they do not say so in the book, it seems they are embarrassed at having raised the issue before. McAfee is quoted as telling an interviewer, "If I had to do it over again, I would put more emphasis on the way technology leads to structural changes in the economy, and less on jobs, jobs, jobs. The central phenomenon is not net job loss. It's the shift in the kinds of jobs that are available."[439]

In 2019, McAfee published "More From Less", a description of how developed economies are consuming less resources while producing more and better goods and services. This thesis encourages my view that the economy of abundance is coming, and may be the way to make technological unemployment a good thing.[440]

Richard and Daniel Susskind

Father and son team Richard and Daniel Susskind published "The Future of the Professions: How Technology Will Transform the Work of Human Experts" in October 2015. Richard Susskind has impressive credentials, having worked in legal technology since the early 1980s, advised numerous government and industry bodies, and garnered a clutch of honorary fellowships from prestigious universities.[441] Perhaps even more impressive is that he seems to have retained the respect of his subjects while lambasting them as inefficient, and doomed to extinction.

The Susskinds describe the "grand bargain" whereby members of the professions (lawyers, doctors, architects, etc.) enjoy a lucrative monopoly over the provision of certain kinds of advice in return for policing the standard of that provision. They argue that this bargain has broken down, with many professional services now being available only to the rich and well-connected. They demonstrate how important this is by illustrating the size of the professions. Healthcare in the US alone now costs $3 trillion a year – more than the GDP of the world's fifth-largest country. The combined revenue of the Big Four accounting firms, at $120 billion, is greater than the GDP of the sixtieth-largest country.

Based on 30 years' experience of the legal industry and backed up by extensive research, the Susskinds paint two scenarios for its mid-term future. The first has professionals working closely with technology, their services enhanced as it improves. The second has most or all of the traditional tasks of professionals carried out by machines. The Susskinds believe that this second outcome is the inevitable one, since what the rest of society cares about is not interaction with humans, but getting our legal, medical and other problems sorted out with the minimum of fuss, risk and expense.

Susskinds keep their focus on the professions, and refrain from making the obvious read-across to the economy as a whole. As a result, they have little to say about universal basic income, or the possibility of society fracturing. But they do note that once machines have taken on responsibility for most or all the tasks previously carried out by human professionals, big questions will be asked about who should own the machines. They don't provide answers to these questions, although they indicate their preference for some form of common ownership which does not involve the state. In this respect they deserve credit for following the logic of their arguments further than most people writing on the subject.

The book is written with refreshing clarity, precision and felicity of expression – and with such a gloomy message for its target audience, that is probably just as well.

In 2019, Daniel Susskind published "A World Without Work", in which he did make the obvious logical step to declaring that technological unemployment is likely across the whole economy. He sees the solution as a big state which implements massive transfers of income and / or wealth, and create a positive outcome. He does not like UBI as a policy, and he does not consider abundance as the way to make the transfers possible.[442]

Scott Santens

Scott Santens is a writer and a campaigner for Universal Basic Income, based in New Orleans.[443] He is a moderator of the Reddit Basic Income page, where he maintains a useful FAQ on the subject.[444] Self-employed since 1997, towards the end of 2015 he managed to procure a basic income for himself based on pledges from others who support his campaign, via the online giving site Patreon.

Jerry Kaplan

Serial entrepreneur Jerry Kaplan co-founded GO Corporation, which was a precursor to smartphones and tablets, and was sold to AT&T. He also co-founded OnSale, an internet auction site which pre-dated Ebay, and was sold for $400m. He teaches at his alma mater, Stanford University, and writes books, including one called "Humans Need Not Apply".

Its message is similar to "The Second Machine Age": AI has reached a tipping point and is becoming powerfully effective. It will disrupt most walks of life, and unless we manage the transition well, the resulting economic instability and growing inequality could be damaging.

Like Ford, Brynjolfsson and McAfee, Kaplan thinks the existing market economy can survive this transition intact.

CGP Grey

Kaplan got the title "Humans Need Not Apply" from a video of the same name[445] which appeared on the internet a year before. Posted to YouTube in August 2014 by an Irish-American who goes by the name CGP Grey (his full name is Colin Gregory Palmer Grey[446]), the video attracted over 5 million views within a year.

The video is well-produced, engaging and persuasive. It contains plenty of technological eye-candy, and makes its points in punchy sound-bites – ideal for today's short attention spans. Unlike the books described above, it offers no solutions to the problems raised by AI and robotic automation, but – also unlike them – it suggests that capitalism cannot cope with what is coming.

Federico Pistono

Federico Pistono is an Italian lecturer and social entrepreneur. He attracted considerable attention with his 2012 book "Robots Will Steal Your Job, But That's OK". A range of eminent people, including Google's Larry Page, were drawn to its optimistic and discursive style.

After making a forceful case that future automation will render most people unemployed, Pistono argues that there is no need to worry. Much of the book is taken up with musing on the nature of happiness – the word features in the titles of a quarter of its chapters. He is hopeful that we will all discover that the pursuit of happiness through material goods is a fool's errand, and he argues that salvation lies in downsizing. He offers the example of his own family, living in northern Italy. They spend $45,000 a year, but by getting rid of two of their three cars, growing their own food, and generating their own electricity, they can reduce this to $29,000 a year.

He also urges us all to educate ourselves – and encourage everyone else to do likewise – but more for personal fulfilment than in a vain attempt to remain employable.

Andy Haldane

As the chief economist of the Bank of England, Andy Haldane isn't the most obvious person to be found musing about the benefits of universal basic income. But that is exactly what he did in a speech at the Trades Union Congress in November 2015[447]. He wondered whether the displacement effect of automation, whereby jobs are destroyed, might start to outweigh the compensation effect, whereby automation raises productivity sufficiently to generate more demand and thus work.

In his speech, Haldane avoided giving a definitive answer to the question of whether we are nearing "peak human", but he raised many of the concerns explored in this book. He presented an estimate prepared by the Bank of England of the likelihood of automation of the jobs in a range of economic sectors in the UK, adapted from the estimates produced for the US by Frey and Osborne. The Bank estimated the UK's situation as slightly less alarming than that of the US, but not much. It found that roughly a third of jobs have a low probability of being automated out of existence, another third have a medium probability, and the final third have a high probability. Haldane avoided putting a specific timescale on this, and also avoided saying what would happen after that undisclosed period.

Martin Wolf

As the main financial columnist and associate editor at the Financial Times, Martin Wolf is the very epitome of a City establishment figure. He was described by US Treasury Secretary Larry Summers as "probably the most deeply thoughtful and professionally informed economic journalist in the world."[448] Although the credit crunch and subsequent recession have re-kindled his youthful enthusiasm for Keynesian economics, it is still a surprise to read him advocating income

420

redistribution and universal basic income, as he did in this article from February 2014:

"If Mr Frey and Prof Osborne [see below] are right [about automation]... we will need to redistribute income and wealth. Such redistribution could take the form of a basic income for every adult, together with funding of education and training at any stage in a person's life. ... The revenue could come from taxes on bads (pollution, for example) or on rents (including land and, above all, intellectual property). Property rights are a social creation. The idea that a small minority should overwhelming[ly] benefit from new technologies should be reconsidered. It would be possible, for example, for the state to obtain an automatic share in the income from the intellectual property it protects."[449]

APPENDIX 2: SKEPTICS

(WRITERS WHO DISMISS THE PROSPECT OF TECHNOLOGICAL UNEMPLOYMENT)

David Autor

David Autor is a professor of economics at MIT. As noted above, he sounded the alarm in a 2003 paper about the "hollowing out" of middle class jobs in the USA – the fact that income has increased for high earners and (albeit less rapidly) for low earners, but stagnated for medium-level earners.

In an interview in October 2015[450], he gave three reasons why he thinks that some observers have been unduly pessimistic, even hysterical, about the likelihood of job destruction. One is that machines complement and augment humans: they always have, and there is no reason to think that is about to change. The second is that machines increase productivity, which creates wealth, consumption; and demand, which creates more jobs.

The third reason is that humans are creative and ingenious. There are many important businesses and activities now that could not have been imagined 50 years ago. In fact, Autor accuses Martin Ford of arrogance in writing off human ingenuity.

In an article for the Journal of Economic Perspectives (summer 2015) entitled "Why are there still so many jobs?",[451] Autor forecasts that people will retain a comparative advantage in so-called "human" attributes such as interpersonal interaction, flexibility, and adaptation. He argues that many jobs – like radiology – combine these attributes with the routine, predictable tasks where computers win. Autor believes it will not be possible to separate these two types of tasks, so the humans cannot be removed.

More generally, Autor is also one of those who believe the rate of change today is over-hyped. He believes that the effect of Moore's Law is substantially muted by regulatory and social frictions which slow down the adoption of new technologies, and he also argues that many technological advances simply don't translate into tangible improvements in the real world. For instance he accepts that his current computer is a thousand times faster than the one he used a few years ago, but he suspects it only makes him 20% more productive. It may be true, he teases, that a new washing machine has more processing power than NASA used to send Neil Armstrong to the moon in 1969, but the washing machine is still not going to the moon.

He derides some of the heralded achievements of AI researchers, arguing for instance that self-driving cars do not emulate human drivers, but instead rely on precise maps of the terrain which have to be prepared before the journey starts. This makes them less flexible than humans, and not fit to be released into the wild without human escorts.

Although Autor is broadly optimistic about our future, he believes that much depends on the decisions that we take. "If machines were in fact to make human labour superfluous, we would have vast aggregate wealth but a serious challenge in determining who owns it and how to share it." He points out that Norway and Saudi Arabia both enjoy economic abundance (thanks to oil rather than AI), but they use it very differently. Norwegians, he says, work few hours per day and are generally happy; Saudis import 90% of their labour and nurture terror.

Robin Hanson

Robin Hanson is an associate professor of economics at George Mason University, in Virginia, USA. Like David Autor, Hanson castigates Martin Ford for inappropriate motives, but whereas Autor accuses Ford of arrogance, Hanson alleges dishonesty: "In the end, it seems that Martin Ford's main issue really is that he dislikes the increase in inequality and wants more taxes to fund a basic income guarantee. All that stuff about robots is a distraction."[452]

After a few more jibes, Hanson addresses Ford's actual thesis. He starts by admitting that "Ford is correct that, … in the long run, robots will eventually get good enough to take pretty much all jobs. But why should we think something like that is about to happen, big and fast, *now*?" He attributes four arguments to Ford, and makes short work of the first three. The first is the Frey and Osborne study, which Hanson dismisses as subjective. The second argument is the decline in labour's share of income since 2000, which Hanson replies could be caused by numerous other factors rather than technological automation. The third argument is the rapid fall in computer prices, which Hanson says has yet to cause any detectable unemployment.

"And then there is Ford's fourth reason: all the impressive computing demos he has seen lately." Hanson is referring, of course, to Google's self-driving cars, real-time machine translation systems, DeepMind's Atari-playing system and so on. Hanson is less impressed by these demonstrations of rapidly improving AI: "We do expect automation to take most jobs eventually, so we should work to better track the situation. But for now, Ford's reading of the omens seems to me little better than fortune telling with entrails or tarot cards."

Having unburdened himself of this cynicism, Hanson proceeds to offer a constructive suggestion. He advocates forecasting by means of prediction markets, where people place bets on particular economic or policy outcomes, like the level of unemployment at some future date. He argues that prediction markets give us a financial stake in being accurate when we make forecasts, rather than just trying to look good to our peers.

Tyler Cowen

A professor at George Mason University and co-author of an extremely popular blog, Tyler Cowen was New Jersey's youngest ever chess champion. He is a man with prodigiously broad knowledge and interests, and although he proposes some key ideas forcefully, there is always some nuance, and he dislikes simplistic and modish solutions. In "The Great Stagnation" (2011) and "Average is Over" (2014), he paints a picture of America's future which is slightly depressing, but not apocalyptic. He is alive to the prospect of dramatically improved AI, and the effect it will have on employment. But he does not think widespread permanent unemployment will be one of its results.

For several years, Cowen has championed the claim that the US economy is hollowing out. He expects automation to continue this trend, perhaps to accelerate it. In an article for Politico magazine[453], he wrote:

"I imagine a world in which, say, 10 to 15 percent of the citizenry … is extremely wealthy and has fantastically comfortable and stimulating lives, equivalent to those of current-day millionaires, albeit with better health care. Much of the rest of the country will have stagnant or maybe even falling wages in dollar terms." This grim outlook for the majority is softened because "they will have a lot more opportunities for cheap fun and cheap education [thanks to] all the free or nearly free services that modern technology makes available." But there is a sting in the tail for the real underclass. They, he says, "will fall by the wayside."

Cowen does not expect a universal basic income to be required. Nor does he expect riots. One reason is that the US population will be older: "By 2030, about 19 percent of the US population will be over 65; in other words, we'll be as old as Floridians are today." Floridians are a conservative bunch, not given to mayhem. Another reason for the absence of riots is that people will increasingly cluster geographically according to income. Few people in the poorer 85% will live in the hothouse cities of San Francisco and New York, and they will not have the wealth of Manhattan waved in their faces. And perhaps most important, the masses will inure themselves with the opiates of free entertainment and social media.

Geoff Colvin

Geoff Colvin is an editor at Fortune magazine. In August 2015 he published "Humans Are Underrated: What High Achievers Know That Brilliant Machines Never Will." It accepts that for the first time, technology may be reducing total employment rather than increasing it, but is skeptical for two reasons in particular. First, Colvin argues that because it is so hard to foresee the new types of jobs that are created when economies shift (just as web development and social media marketing were hard to foresee), we under-estimate how many of them there will be.

Second, Colvin believes that skills of deep human interaction – empathy, storytelling, the ability to build relationships - will become far more valuable in the future, and many people will be able to prosper by bringing those skills into the evolving economy. "We're hard-wired by 100,000 years of evolution to value deep interaction with other humans (and not with computers). Those wants won't be changing anytime soon."[454]

Roger Bootle

In his 2019 book, "The AI Economy", the noted economist Roger Bootle dismisses the possibility of technological unemployment from the outset, and he makes little effort to hide his disdain for those who take the idea seriously. He labels me an "AI visionary", a "geek", a "bubbling enthusiast", but also a pessimist, "emanating gloom".[455] Ouch.

He thinks "there are many areas where humans possess an absolute advantage over robots and AI, including manual dexterity, emotional intelligence, creativity, flexibility, and most importantly, humanity." And he thinks this situation will last forever. In reality, the human advantages in all those areas are gone or going, except for humanity, which is a category machines will probably never join. It is by no means clear that humanity is a characteristic which will allow us to get and keep paid jobs forever. Or, happily, require us to.

ACKNOWLEDGE-MENTS

I am enormously grateful to the following alphabetised people, who have given their time and energy to support this book, and in many cases provided constructive criticism on its earlier drafts. I have learned a lot from their insights, and the book is much better for them. All errors, omissions and solecisms are of course my fault, not theirs.

Adam Jolly, Alexander Chace, Aubrey de Grey, Ben Goertzel, Ben Goldsmith, Ben Medlock, Brad Feld, Carl Frey, Charles Radclyffe, Chris Meyer, Dan Goodman, Daniel Hulme, Daniel van Leeuwen, David Fitt, David Shukman, David Wood, Ed Charvet, Gerald Huff, Henry Whitaker, Hugh Pym, Hugo de Garis, Jaan Tallinn, James Hughes, Janet Standen, Jeff Pinsker, Jim Muttram, Joe Pinsker, John Danaher, John Higgins, Justin Stewart, Keith Bullock, Kenneth Cukier, Leah Eatwell, Lesley Fenton O'Creevy, Malcolm Myers, Mark Chadwick, Mark Mardell, Mark Mathias,

Mike Johnston, Olly Buston, Parry Hughes-Morgan, Peter Fenton O'Creevy, Peter Monk, Rachel Lawston, Radhika Chadwick, Randal Koene, Ray Eitel-Porter, Russell Buckley, Scott Heekin-Canedy, Shamus Rae, Simon Thorpe, Stuart Armstrong, Stuart Russell, Tess Read, Tim Cross, Tom Aubrey, Tom Hunter, Will Gilpin, William Charlwood, William Graham.

The hugely talented book designer Rachel Lawston has produced a cover which (I think) looks great and answers the brief perfectly.

Last but definitely not least, my profoundest thanks go to my partner Julia, who is my adviser, my cheerleader, and my kindest but most penetrating critic.

ENDNOTES

2   The term economic singularity was first used
(as far as I can tell) by the economist Robin Hanson.
He uses the phrase in a different sense, to mean a
time of very high growth in per-capita consumption:
http://mason.gmu.edu/~rhanson/fastgrow.html

3   http://arxiv.org/pdf/0712.3329v1.pdf

4   http://skyview.vansd.org/lschmidt/Projects/
The%20Nine%20Types%20of%20Intelligence.htm

5   http://www.savethechimps.org/about-us/chimp-
facts/

6   The term AGI has been popularised by AI
researcher Ben Goertzel, although he gives credit for
its invention to Shane Legg and others: http://wp.go-
ertzel.org/who-coined-the-term-agi/

7   https://singularityhub.com/2017/03/29/
google-chases-general-intelligence-with-new-

ai-that-has-a-memory/?utm_content=buff-er958e8&utm_medium=social&utm_source=twit-ter-hub&utm_campaign=buffer

8   http://www.etymonline.com/index.php?ter-m=algorithm

9   The Shape of Automation for Men and Management by Herbert Simon, 1965

10   Computation: Finite and Infinite Machines by Marvin Minsky, 1967

11   https://www.forbes.com/sites/gil-press/2020/03/30/12-artificial-intelligence-ai-mile-stones-3-computer-graphics-give-birth-to-big-data/#7141122a21cf

12   https://setandbma.wordpress.com/2013/02/04/who-coined-the-term-big-data/

13   http://www.iflscience.com/technology/how-much-data-does-the-world-generate-every-minute/

14   https://en.wikipedia.org/wiki/People%27s_Lib-eration_Army#Third_Department

15   https://www.youtube.com/watch?v=G_HgfrD5t-DQ

16   https://datascience.stackexchange.com/ques-

tions/37078/source-of-arthur-samuels-defini-
tion-of-machine-learning

17 http://www.wired.com/2016/01/microsoft-neu-
ral-net-shows-deep-learning-can-get-way-deeper/

18 https://www.technologyreview.com/s/608911/
is-ai-riding-a-one-trick-pony/

This version of events is strenuously disputed by
another veteran of AI research, Jurgen Schmidhuber,
who claims the group in Toronto claim credit for
innovations achieved previously and elsewhere. They
in turn reject his allegations. http://people.idsia.
ch/~juergen/critique-honda-prize-hinton.html

19 https://bernardmarr.com/default.asp?conten-
tID=1902

20 https://medium.com/ai%C2%B3-theory-prac-
tice-business/understanding-hintons-capsule-net-
works-part-i-intuition-b4b559d1159b

21 http://www.computerweekly.com/feature/Apol-
lo-11-The-computers-that-put-man-on-the-moon

22 http://www.bradford-delong.com/2017/09/
do-they-really-say-technological-progress-is-slow-
ing-down.html

23 https://www.internetlivestats.com/one-sec-

ond/#google-band If that makes you curious to know about your own Google searches, you can find them here: google.com/history

24   https://seekingalpha.com/article/4230982-algo-trading-dominates-80-of-stock-market

25   The Big Switch by Nicholas Carr (p 212)

26   https://www.crunchbase.com/organization/google/acquisitions/acquisitions_list

27   https://www.wired.com/2017/02/inside-facebooks-ai-machine/

28   http://www.theguardian.com/technology/2016/jan/31/viv-artificial-intelligence-wants-to-run-your-life-siri-personal-assistants

29   https://techcrunch.com/2017/07/19/apple-launches-machine-learning-research-site/

30   https://techcrunch.com/2016/09/29/microsoft-forms-new-ai-research-group-led-by-harry-shum/

31   https://www.geekwire.com/2017/one-year-later-microsoft-ai-research-grows-8k-people-massive-bet-artificial-intelligence/

32   https://www.cnbc.com/2019/09/28/amazon-al-

exa-growth-has-investors-questioning-the-busi-ness-model.html

33   http://www.aaai.org/Magazine/Watson/watson.php

34   http://www.latimes.com/business/technology/la-fi-cutting-edge-ibm-20160422-story.html

35   https://spectrum.ieee.org/biomedical/diagnos-tics/how-ibm-watson-overpromised-and-underde-livered-on-ai-health-care

36   https://en.wikipedia.org/wiki/List_of_most_popular_websites

37   https://venturebeat.com/2019/12/31/baidu-se-cures-licenses-to-test-self-driving-cars-in-beijing/

38   https://www.forbes.com/sites/bernard-marr/2018/06/04/artificial-intelligence-ai-in-chi-na-the-amazing-ways-tencent-is-driving-its-adop-tion/#68266017479a

39   https://www.washingtonpost.com/news/the-switch/wp/2016/10/13/china-has-now-eclipsed-us-in-ai-research/?utm_term=.d4f9eb1c9353 This article is behind a paywall. An alternative source is here: https://futurism.com/china-has-overtaken-the-u-s-in-ai-research/

40   http://thediplomat.com/2017/07/chinas-artifi-cial-intelligence-revolution/

41   http://www.shanghaidaily.com/metro/society/Shanghai-will-be-at-the-heart-of-Chinas-artificial-intelligence/shdaily.shtml

42   http://www.wired.com/2015/11/goo-gle-open-sources-its-artificial-intelligence-engine/

43   https://www.theguardian.com/technology/2016/apr/13/google-updates-tensorflow-open-source-arti-ficial-intelligence

44   http://www.wired.com/2015/12/facebook-open-source-ai-big-sur/

45   https://www.technologyreview.com/2020/06/22/1004251/baidus-deep-learning-platform-fuels-the-rise-of-industrial-ai/

46   https://bernardmarr.com/default.asp?conten-tID=1171

47   https://venturebeat.com/2020/01/22/cb-in-sights-ai-startup-funding-hit-new-high-of-26-6-bil-lion-in-2019/

48   https://www.youtube.com/watch?v=Skfw282fJak

49   http://futureoflife.org/2016/01/27/are-humans-

dethroned-in-go-ai-experts-weigh-in/

50   https://www.theguardian.com/technology/2017/
dec/07/alphazero-google-deepmind-ai-beats-cham-
pion-program-teaching-itself-to-play-four-hours

51   https://www.theverge.com/2017/8/11/16137388/
dota-2-dendi-open-ai-elon-musk

52   https://www.technologyreview.com/s/614650/ai-
deepmind-outcompeted-most-players-at-starcraft-ii/

53   http://www.popsci.com/scitech/article/2004-06/
darpa-grand-challenge-2004darpas-debacle-desert

54   http://www.pcmag.com/arti-
cle2/0,2817,2370598,00.asp

55   https://venturebeat.com/2020/01/06/waymos-
autonomous-cars-have-driven-20-million-miles-on-
public-roads/

56   https://www.cbinsights.com/research/autono-
mous-driverless-vehicles-corporations-list/

57   https://www.cnet.com/roadshow/news/
waymo-apple-av-disengagement-reports-califor-
nia-2018/

58   https://www.cnbc.com/2019/11/19/reuters-
america-update-2-u-s-safety-board-chair-criticizes-

uber-for-2018-fatal-self-driving-crash.html

59 http://www.thechurchofgoogle.org/Scripture/
Proof_Google_Is_God.html

60 https://www.reddit.com/r/churchofgoogle/

61 https://techcrunch.com/2019/06/12/uber-airs-
plan-to-get-you-from-a-skyport-to-an-airport/

62 https://www.theverge.com/2020/2/4/21122341/
wisk-flying-taxi-kitty-hawk-boeing-cora-new-zea-
land

63 http://www.thedrum.com/opinion/2016/02/08/
why-artificial-intelligence-key-google-s-battle-ama-
zon

64 http://www.wired.com/2012/06/google-x-neu-
ral-network/

65 They are the Pembroke and the Cardigan Corgi.
http://research.microsoft.com/en-us/news/features/
dnnvision-071414.aspx

66 http://news.sciencemag.org/social-scienc-
es/2015/02/facebook-will-soon-be-able-id-you-any-
photo

67 http://www.computerworld.com/arti-
cle/2941415/data-privacy/is-facial-recognition-a-

threat-on-facebook-and-google.html

68   https://www.gemalto.com/govt/biometrics/facial-recognition

69   https://cloud.google.com/vision/

70   https://nymag.com/intelligencer/2019/11/the-future-of-facial-recognition-in-america.html

71   http://www.bloomberg.com/news/2014-12-23/speech-recognition-better-than-a-human-s-exists-you-just-can-t-use-it-yet.html

72   http://www.businessinsider.fr/uk/microsofts-speech-recognition-5-1-error-rate-human-level-accuracy-2017-8/

73   http://www.forbes.com/sites/parmyolson/2014/05/28/microsoft-unveils-near-real-time-language-translation-for-skype/

74   https://www.goodreads.com/quotes/1187961-the-babel-fish-is-small-yellow-and-leech-like-and-probably

75   https://www.cnet.com/how-to/pixel-buds-2-the-lowdown-on-googles-new-floating-wireless-earbuds-for-2020/

76   http://www.zdnet.com/article/lip-reading-bots-

in-the-wild/

77   https://towardsdatascience.com/bert-explained-state-of-the-art-language-model-for-nlp-f8b21a9b6270

78   https://www.zdnet.com/article/openais-gigantic-gpt-3-hints-at-the-limits-of-language-models-for-ai/

79   https://youtu.be/V1eYniJ0Rnk?t=1

80   http://edge.org/response-detail/26780

81   https://www.theguardian.com/technology/2017/aug/06/artificial-intelligence-and-will-we-be-slaves-to-the-algorithm

82   https://www.theguardian.com/technology/2015/jun/18/google-image-recognition-neural-network-androids-dream-electric-sheep

83   https://www.youtube.com/watch?v=deyOX6Mt As

84   http://www.uh.edu/engines/epi265.htm

85   http://www.bbc.co.uk/news/technology-35977315

86   Moravec wrote about this phenomenon in his 1988 book "Mind Children". A possible explanation

is that the sensory motor skills and spatial awareness that we develop as children are the product of millions of years of evolution. Rational thought is something we have only been doing for a few thousand years. Perhaps it really isn't hard, but just seems hard because we are not yet optimised for it.

87   http://www.tomdispatch.com/post/175822/tom-gram%3A_crump_and_harwood%2C_the_net_closes_around_us/

88   http://www.newyorker.com/tech/elements/little-brother-is-watching-you

89   http://www.wired.com/2014/03/going-tracked-heres-way-embrace-surveillance/

90   https://www.washingtonpost.com/news/the-switch/wp/2016/03/28/mass-surveillance-silences-minority-opinions-according-to-study/

91   http://www.bbc.co.uk/news/world-asia-china-34592186

92   http://www.theguardian.com/technology/2015/oct/06/peeple-ratings-app-removes-contentious-features-boring

93   https://singularityhub.com/2017/09/25/will-privacy-survive-the-future/#.WcrRq8lOYKY.twitter

94 https://www.technologyreview.com/s/601294/microsoft-and-google-want-to-let-artificial-intelligence-loose-on-our-most-private-data/?utm_source=Twitter&utm_medium=tweet&utm_campaign=@KyleSGibson

95 https://solidproject.org/

96 https://www.technologyreview.com/s/604122/the-financial-world-wants-to-open-ais-black-boxes/?utm_campaign=add_this&utm_source=twitter&utm_medium=post

97 http://www.wired.co.uk/article/machine-learning-bias-prejudice

98 https://www.forbes.com/sites/cognitive-world/2018/09/07/shooting-the-messenger/#30167d4d6584

99 https://wt.social/

100 https://www.youtube.com/watch?v=Hip-TO_7mUOw

101 https://www.washingtonpost.com/local/public-safety/the-new-way-police-are-surveilling-you-calculating-your-threat-score/2016/01/10/e42bccac-8e15-11e5-baf4-bdf37355da0c_story.html

102 https://www.wired.com/story/when-govern-

ment-rules-by-software-citizens-are-left-in-the-dark/

103 https://www.theguardian.com/technology/2017/sep/07/new-artificial-intelligence-can-tell-whether-youre-gay-or-straight-from-a-photograph

104 The Economist, December 4, 2003

105 Douglas Adams, The Salmon of Doubt

106 http://www.wired.com/2014/10/future-of-artificial-intelligence/

107 https://www.gsb.stanford.edu/insights/andrew-ng-why-ai-new-electricity

108 https://www.pocket-lint.com/phones/news/google/146008-what-is-google-duplex-where-is-it-available-and-how-does-it-work

109 https://www.theverge.com/2019/11/21/20975599/google-assistant-duplex-cinema-movie-tickets-usa-uk-android

110 https://ai.googleblog.com/2020/01/towards-conversational-agent-that-can.html

111 Not an everyday object outside the USA, of course

112 http://www.bloomberg.com/news/articles/2016-01-11/google-chairman-thinks-ai-can-help-solve-world-s-hard-problems-

113 An ocarina is a wind instrument about the size of a fist. First introduced to Europeans by the Aztecs, it looks like a toy submarine.

114 https://www.extremetech.com/extreme/171992-motorola-patents-e-tattoo-that-can-read-your-thoughts-by-listening-to-unvocalized-words-in-your-throat

115 This is actually a great idea, which is being trialled in Argentina at the time of writing: http://www.telegraph.co.uk/motoring/motoringvideo/11680348/Transparent-trucks-with-rear-mounted-Samsung-safety-screens-set-to-save-overtaking-drivers.html Of course it may be less valuable when cars drive themselves and their human occupants don't look at the road.

116 https://www.who.int/violence_injury_prevention/road_safety_status/2018/en/

117 http://www.japantimes.co.jp/news/2015/11/15/business/tech/human-drivers-biggest-threat-developing-self-driving-cars/#.Vo7D5fmLRD8

118 http://www.theatlantic.com/business/archive/2013/02/the-american-commuter-spends-38-

hours-a-year-stuck-in-traffic/272905/

119 http://www.reinventingparking.org/2013/02/
cars-are-parked-95-of-time-lets-check.html

120 Sorry, couldn't resist the nostalgic reference to
an old ad: https://www.youtube.com/watch?v=J-
dA9CorwMiU

121 http://www.etymonline.com/index.php?ter-
m=autocar

122 http://www.digitaltrends.com/cars/audi-autono-
mous-car-prototype-starts-550-mile-trip-to-ces/

123 http://www.reuters.com/investigates/special-re-
port/autos-driverless/

124 https://techcrunch.com/2020/03/02/waymo-
brings-in-2-25-billion-from-outside-investors-al-
phabet/

125 It has been suggested that electric cars should
make noises so that people don't step off the pave-
ment in front of them. A friend tells me he would
like his to make a noise like two coconuts being
banged together, in homage to the opening scene
in Monty Python and the Holy Grail, where King
Arthur, unable to afford a horse, has a camp follower
fake the noise of one with coconuts.

126 https://www.forbes.com/sites/bradtempleton/2019/02/21/robocar-engineers-prefer-to-solve-the-runaway-trolley-problem-by-fixing-the-brakes on-the-trolley/#5bcb6e9d36fc

127 http://www.rfidjournal.com/articles/view?4986

128 http://www.vdi-nachrichten.com/Technik-Gesellschaft/Industrie-40-Mit-Internet-Dinge-Weg-4-industriellen-Revolution

129 Coined by another British entrepreneur, Simon Birrell: https://www.linkedin.com/in/simonbirrell

130 https://www.gartner.com/en/newsroom/press-releases/2019-08-29-gartner-says-5-8-billion-enterprise-and-automotive-io

131 https://www.helpnetsecurity.com/2019/06/21/connected-iot-devices-forecast/

132 http://singularityhub.com/2016/02/09/when-the-world-is-wired-the-magic-of-the-internet-of-everything/

133 http://www.telegraph.co.uk/technology/internet/12050185/Marc-Andreessen-In-20-years-every-physical-item-will-have-a-chip-implanted-in-it.html

134 http://www.information-age.com/it-management/strategy-and-innovation/123460379/

trains-brains-how-artificial-intelligence-transform-
ing-railway-industry

135 http://home.cern/topics/birth-web

136 http://www.abc.net.au/news/2017-08-14/
how-ai-could-put-an-end-to-prisons-as-we-know-
them/8794910

137 https://ifr.org/downloads/press2018/IFR
World Robotics Outlook 2019 - Chicago.pdf

138 https://www.wired.com/story/the-education-of-
brett-the-robot/

139 https://www.engadget.com/2019-04-23-laun-
droid-robot-seven-dreamers-bankruptcy.html

140 http://www.nextgov.com/emerg-
ing-tech/2016/05/robots-are-starting-learn-
touch/128065/

141 https://www.recode.net/2017/6/8/15766440/
softbank-alphabet-google-robotics-boston-dynam-
ics-schaft

142 http://uk.businessinsider.com/softbank-ceo-ma-
sayoshi-son-thinks-singularity-will-occur-with-
in-30-years-2017-2

143 https://www.washingtonpost.com/gd-

pr-consent/?next_url=https%3a%2f%2f-www.washingtonpost.com%2ftechnolo-gy%2f2019%2f06%2f07%2famazons-one-day_deliv-ery-service-depends-work-thousands-robots

144 http://www.theguardian.com/world/2015/sep/28/no-sex-with-robots-says-japanese-android-firm-softbank

145 https://www.theguardian.com/technology/2015/aug/03/hitchbot-hitchhiking-robot-destroyed-phila-delphia

146 http://www.telegraph.co.uk/news/science/science-news/12073587/Meet-Nadine-the-worlds-most-human-like-robot.html

147 https://spectrum.ieee.org/automaton/ro-botics/humanoids/an-uncanny-mind-masahi-ro-mori-on-the-uncanny-valley

148 http://techcrunch.com/2016/01/07/the-grillbot-is-a-robot-that-cleans-your-grill/#.w9z87m:Hd0d

149 http://intl.eksobionics.com/

150 http://singularityhub.com/2016/02/29/drones-have-reached-a-tipping-point-heres-what-happens-next/

151 https://www.technologyreview.com/the-down-

load/608912/urban-drone-deliveries-are-finally-tak-ing-flight/

152 https://www.technologyreview.com/the-download/608718/finally-theres-a-halfway-compel-ling-consumer-drone-delivery-service/

153 https://newatlas.com/amazon-new-deliv-ery-drones-us-faa-approval/36957/

154 https://www.komando.com/online-shopping/look-up-in-the-sky-its-my-package-amazon-to-start-drone-delivery-within-months/571255/

155 I was one of them! https://www.nytimes.com/2018/12/27/world/europe/gatwick-airport-drone.html

156 https://en.wikipedia.org/wiki/List_of_the_busi-est_airports_in_Europe

157 https://www.thebureauinvestigates.com/drone-war/data/the-bush-years-pakistan-strikes-2004-2009

158 https://www.economist.com/unit-ed-states/2014/06/23/dilbert-at-war

159 https://www.documentcloud.org/docu-ments/5985986-annual-report-civilian-casual-ties-in-connection.html

160 https://web.archive.org/web/20110830213657/
http://counterterrorism.newamerica.net/drones

161 https://www.wired.com/2015/05/try-goo-
gle-cardboard/

162 https://www.theverge.com/2019/11/6/20952495/
google-cardboard-open-source-phone-based-vr-
daydream

163 https://venturebeat.com/2019/10/15/google-dis-
continues-daydream-vr/

164 Your brain is wired to make you see things
before you hear them as it knows that light travels
faster than sound. Thus the brain can tolerate audio
lagging video, but is much less tolerant of video
lagging audio. This is known as Multi-Modal Perception.

165 https://www.extremetech.com/gaming/245262-
facebook-slashes-oculus-rift-prices-user-growth-
sags

166 https://www.polygon.com/2017/2/27/14753570/
pokemon-go-downloads-650-million

167 https://www.cnbc.com/2017/09/08/apples-arkit-
will-bring-with-it-a-new-form-of-mobile-advertis-
ing.html

168 http://variety.com/2017/digital/news/magic-leap-funding-temasek-1202559027/

169 https://www.cnet.com/news/apple-glasses-leaks-and-rumors-when-could-smartglasses-arrive/

170 https://techcrunch.com/2020/06/30/google-acquires-smart-glasses-company-north-whose-focals-2-0-wont-ship/

171 https://www.mic.com/p/second-life-still-has-dedicated-users-in-2020-heres-what-keeps-them-sticking-around-18693758

172 https://www.cnet.com/news/i-tried-facebooks-vision-for-the-social-future-of-vr-full-of-question-marks/

173 https://www.viar360.com/virtual-reality-market-size-2018/

174 The games industry is much bigger than Hollywood if you stop measuring move income at the box office. If you add in DVD and other "windows", plus merchandising, it is hard to say. https://www.quora.com/Who-makes-more-money-Hollywood-or-the-video-game-industry

175 https://versions.killscreen.com/we-should-be-talking-about-torture-in-vr/

176 https://www.macrotrends.net/stocks/charts/KODK/eastman-kodak/total-assets

177 https://www.statista.com/statistics/339845/company-value-and-equity-funding-of-airbnb/

178 https://www.macrotrends.net/stocks/charts/H/hyatt-hotels/market-cap

179 https://www.statista.com/statistics/824612/number-of-hyatt-hotels-worldwide/

180 https://about.hyatt.com/en.html

181 https://www.macrotrends.net/stocks/charts/H/hyatt-hotels/revenue

182 https://www.owler.com/company/airbnb

183 https://skift.com/2019/09/19/airbnb-quarterly-revenue-tops-1-billion-for-the-second-time/

184 A term invented at defense contractor Lockheed Martin: https://www.lockheedmartin.com/en-us/who-we-are/business-areas/aeronautics/skunkworks.htmlindex.html

185 http://lazooz.org/

186 I am grateful to Russell Buckley for introducing me to this illustration.

187 http://www.pcmag.com/encyclopedia/term/37701/amara-s-law

188 https://quoteinvestigator.com/2019/01/03/estimate/

189 http://www.rationaloptimist.com/blog/amaras-law/

190 https://spectrum.ieee.org/tech-talk/semiconductors/devices/the-murky-origins-of-moores-law

191 http://fortune.com/2017/01/05/intel-ces-2017-moore-law/

192 https://www.economist.com/node/21693710/sites/all/modules/custom/ec_essay

193 Clock speed, also known as clock rate or processor speed, is the number of cycles a chip (central processing unit, or CPU) performs each second. Inside each chip is a small quartz crystal which vibrates, or oscillates, at a particular frequency. It takes a fixed number of oscillations, or cycles, to perform the instructions that a chip is given. One cycle is one Hertz, and today's chips operate in gigaHertz (GHz), billions of cycles per second. As other aspects of chip designs diverge, clock speed is no longer a reliable measure of a chip's effective performance.

194 http://www.extremetech.com/extreme/225353-intel-formally-kills-its-tick-tock-approach-to-processor-development

195 https://www.tsmc.com/tsmcdotcom/PRListing-NewsAction.do?action=detail&language=E&news-id=THPGWQTHTH

196 https://techcrunch.com/2017/05/17/google-announces-second-generation-of-tensor-processing-unit-chips/

197 http://www.popularmechanics.com/technology/a18493/stanford-3d-computer-chip-improves-performance/

198 http://www.engadget.com/2016/03/28/ibm-resistive-processing-deep-learning/

199 https://newsroom.intel.com/editorials/intels-new-self-learning-chip-promises-accelerate-artificial-intelligence/

200 https://www.engineersaustralia.org.au/event/2019/09/brainchips-akida-artificial-intelligence-neuromorphic-system-chip

201 https://www.americanscientist.org/article/is-quantum-computing-a-cybersecurity-threat

202 https://www.ibm.com/blogs/research/2019/10/

on-quantum-supremacy/

203 https://www.forbes.com/sites/moorin-sights/2020/04/30/googles-top-quantum-scientist-explains-in-detail-why-he-resigned/#68e1376b6983

204 https://www.microsoft.com/en-us/research/blog/machine-learning-and-the-learning-machine-with-dr-christopher-bishop/

205 https://www.weforum.org/pages/the-fourth-in-dustrial-revolution-by-klaus-schwab

206 http://www.slate.com/articles/technology/fu-ture_tense/2016/01/the_world_economic_forum_is_wrong_this_isn_t_the_fourth_industrial_revolu-tion.html

207 The descriptions of these revolutions owes something to Yuval Harari's book Sapiens.

208 Ulam, Stanislaw (May 1958). "Tribute to John von Neumann". 64, #3, part 2. Bulletin of the Amer-ican Mathematical Society: 5. https://docs.google.com/file/d/0B-5-JeCa2Z7hbWcxTGsyU09HSTg/edit?pli=1

209 http://edoras.sdsu.edu/~vinge/misc/singularity.html

210 http://www.kurzweilai.net/images/How-My-

Predictions-Are-Faring.pdf

211 http://spectrum.ieee.org/computing/software/
ray-kurzweils-slippery-futurism

212 https://www.youtube.com/watch?v=aboZctrH-
fK8 Also, try typing "the answer to life the universe
and everything" into Google Search.

213 http://yudkowsky.net/singularity/schools/

214 https://www.scientificamerican.com/article/
when-will-speech-recognition-software-finally-be-
good-enough/

215 https://www.minnpost.com/macro-micro-min-
nesota/2012/02/history-lessons-understanding-de-
cline-manufacturing

216 http://blogs.rmg.co.uk/longitude/2014/07/30/
guest-post-pirate-map/

217 https://www.bls.gov/opub/mlr/1981/11/art2full.
pdf

218 https://www.bls.gov/emp/ep_table_201.htm

219 http://answers.google.com/answers/thread-
view?id=144565

220 http://www.americanequestrian.com/pdf/

AQHA%202012%20Horse%20Statistics.pdf

221 https://en.wikipedia.org/wiki/Automation#cite_note-7

222 M. A. Laughton, D. J. Warne (ed), Electrical Engineer's Reference book

223 http://www.oleantimesherald.com/news/did-you-know-gas-pump-shut-off-valve-was-invented/article_c7a00da2-b3eb-54e1-9c8d-ee36483a7e33.html

224 http://www.ehow.com/about_4678910_ro-bots-car-manufacturing.html

225 https://ifr.org/img/uploads/Executive_Summa-ry_WR_Industrial_Robots_20161.pdf

226 https://www.therobotreport.com/hahn-group-acquires-rethink-robotics-ip/

227 https://www.washingtonpost.com/gd-pr-consent/?next_url=https%3a%2f%2f-www.washingtonpost.com%2ftechnolo-gy%2f2019%2f06%2f07%2famazons-one-day-deliv-ery-service-depends-work-thousands-robots

228 http://www.npr.org/sections/mon-ey/2015/02/05/382664837/map-the-most-common-job-in-every-state

229 http://www.nber.org/papers/w23285

230 https://www.technologyreview.com/s/604005/actually-steve-mnuchin-robots-have-already-affected-the-us-labor-market/

231 http://www.nationalarchives.gov.uk/education/politics/g5/

232 http://jetpress.org/v24/campa2.htm

233 Ricardo originally thought that innovation benefited everyone, but he was persuaded by Malthus that it could suppress wages and cause long-term unemployment. He added a chapter called "On Machinery" to the final edition of his book "On the Principles of Political Economy and Taxation".

234 http://www.theguardian.com/business/2015/aug/17/technology-created-more-jobs-than-destroyed-140-years-data-census

235 https://en.wikipedia.org/wiki/Bowley%27s_law

236 http://www.economics.ox.ac.uk/Department-of-Economics-Discussion-Paper-Series/engels-pause-a-pessimist-s-guide-to-the-british-industrial-revolution

237 https://www.ted.com/talks/david_autor_why_

are there still so many jobs

238 https://www.forbes.com/sites/eriksher-man/2016/12/17/automation-has-created-more-jobs-in-the-past-but-will-it-now/#6ccaf5257cde

239 https://www.bba.org.uk/wp-content/up-loads/2014/08/2014 UK Banking Industry Struc-ture Abstract.pdf

240 Assuming the work is happening on Earth. Wikipedia offers a more general but less euphonious definition: "Work is the product of the force applied and the displacement of the point where the force is applied in the direction of the force."

241 http://www.mckinsey.com/insights/business technology/four fundamentals of workplace auto-mation

242 2013 data: http://www.ons.gov.uk/ons/dcp171778 315661.pdf

243 https://slate.com/business/2012/05/blue-collar-white-collar-why-do-we-use-these-terms.html

244 https://www.invespcro.com/blog/global-on-line-retail-spending-statistics-and-trends/

245 https://www.smallbizlabs.com/2018/08/gallup-says-36-of-us-workers-are-in-the-gig-economy.html

246 https://www.census.gov/library/stories/2019/06/america-keeps-on-trucking.html#:~:text=More%20than%203.5%20million%20people,occupations%20in%20the%20United%20States.

247 http://www.bls.gov/ooh/transportation-and-material-moving/bus-drivers.htm

248 http://www.bls.gov/ooh/transportation-and-material-moving/taxi-drivers-and-chauffeurs.htm

249 http://bbc.in/2uTK2PD

250 http://www.abc.net.au/news/2015-10-18/rio-tinto-opens-worlds-first-automated-mine/6863814

251 http://www.mining.com/why-western-australia-became-the-center-of-mine-automation/

252 Three minutes into the eighth video here: https://a16z.com/2018/02/03/autonomy-ecosystem-frank-chen-summit/

253 https://www.theverge.com/2020/2/11/21133389/house-energy-commerce-self-driving-car-hearing-bill-2020

254 http://www.wsj.com/articles/SB10001424053111903480904576512250915629460

255 http://fortune.com/2014/05/04/6-things-i-learned-at-buffetts-annual-meeting/

256 http://www.thenewspaper.com/news/43/4341.asp

257 http://www.alltrucking.com/faq/truck-drivers-in-the-usa/

258 http://www.npr.org/sections/money/2015/02/05/382664837/map-the-most-common-job-in-every-state

259 http://www.huffingtonpost.com/entry/pittsburgh-uber-self-driving-cars_us_59caa91ae4b0d0b-254c4fcdf?ncid=inblnkushpmg00000009

260 The Programme was established in January 2015 with funding from Citibank, one of the largest financial institutions in the world. The Oxford Martin school was set up as part of Oxford University in 2005, as an institution dedicated to understanding the threats and opportunities facing humanity in the 21st century. It is named after James Martin, a writer, consultant and entrepreneur, who founded the school with what was at the time the largest donation ever made to the university – no mean feat, given that Oxford was founded 1,000 years ago, and is the second oldest university in the world (after Bologna in Italy).

261 http://www.oxfordmartin.ox.ac.uk/downloads/academic/The_Future_of_Employment.pdf

262 https://www.ndtv.com/world-news/face-scans-robot-baggage-handlers-airports-of-the-future-1742507

263 https://www.technologyreview.com/s/608811/drones-and-robots-are-taking-over-industrial-inspection/

264 https://www.mckinsey.com/industries/retail/our-insights/automation-in-retail-an-executive-overview-for-getting-ready#

265 https://wikivisually.com/wiki/Amazon_Go

266 https://en.wikipedia.org/wiki/Horn_%26_Hardart#Automated_food

267 https://gizmodo.com/why-fast-food-is-the-ticking-time-bomb-of-job-automatio-1837898231

268 https://emerj.com/ai-sector-overviews/fast-food-robots-kiosks-and-ai-use-cases/

269 http://blogs.forrester.com/andy_hoar/15-04-14-death_of_a_b2b_salesman

270 https://www.ons.gov.uk/peoplepopulationandcommunity/populationandmigration/international-

migration/articles/labourintheagricultureindustry/2018-02-06

271 https://eandt.theiet.org/content/articles/2019/06/strawberry-picking-robots-to-gather-fruit-for-wimbledon-fans/

272 https://eandt.theiet.org/content/articles/2019/06/strawberry-picking-robots-to-gather-fruit-for-wimbledon-fans/

273 https://spectrum.ieee.org/automaton/robotics/industrial-robots/autonomous-robots-plant-tend-and-harvest-entire-crop-of-barley

274 https://www.agriland.co.uk/farming-news/hands-free-hectare-to-expand-to-35ha-site/

275 https://www.technologyreview.com/s/601215/china-is-building-a-robot-army-of-model-workers/

276 https://www.invespcro.com/blog/global-online-retail-spending-statistics-and-trends/

277 https://www.ons.gov.uk/businessindustryandtrade/retailindustry/timeseries/j4mc/drsi

278 https://www.wired.com/story/grasping-robots-compete-to-rule-amazons-warehouses/

279 https://www.cnbc.com/2017/09/11/new-york-

times-digital-as-amazon-pushes-forward-with-ro-bots-workers-find-new-roles.html

280 https://techcrunch.com/2019/06/05/amazon-says-it-has-deployed-more-than-200000-robotic-drives-globally/

281 https://slate.com/technology/2014/03/quakebot-los-angeles-times-robot-journalist-writes-article-on-la-earthquake.html

282 http://www.theguardian.com/technology/2014/sep/12/artificial-intelligence-data-journalism-media

283 https://www.youtube.com/watch?v=HXKDn-qM9Ulw

284 http://www.chinadaily.com.cn/chi-na/2015-12/24/content_22794242.htm

285 https://medium.com/speaking-naturally/tales-from-the-worlds-most-used-chatbot-microsoft-xiao-ice-s-lead-scientist-on-cutting-edge-acf34fe9a624

286 http://linkis.com/www.theatlantic.com/SoE5e

287 http://persado.com/

288 http://www.techtimes.com/arti-cles/127526/20160126/ai-politics-how-an-arti-ficial-intelligence-algorithm-can-write-politi-

cal-speeches.htm

289 https://www.research.ibm.com/artificial-intelligence/project-debater/about/

290 https://www.ncbi.nlm.nih.gov/pmc/articles/PMC6449411/

291 https://www.statnews.com/2019/10/23/advancing-ai-health-care-trust/

292 https://towardsdatascience.com/why-ai-will-not-replace-radiologists-c7736f2c7d80

293 http://uk.businessinsider.com/deepmind-cofounders-invest-in-babylon-health-2016-1

294 http://forbesindia.com/article/hidden-gems/thyrocare-technologies-testing-new-waters-in-medical-diagnostics/41051/1

295 http://www.wsj.com/articles/SB1000142405270230398390457909325257381413

296 http://www.outpatientsurgery.net/outpatient-surgery-news-and-trends/general-surgical-news-and-reports/ethicon-pulling-sedasys-anesthesia-system--03-10-16

297 http://www.wired.co.uk/news/archive/2016-05/05/autonomous-robot-surgeon

298 http://www.scmp.com/news/china/article/2112197/chinese-robot-dentist_first-fit-implants-patients-mouth-without-any-human?utm_source=t.co&utm_medium=referral

299 http://www.thelancet.com/journals/lanonc/article/PIIS1470-2045(17)30572-7/fulltext?elsca1=tlpr

300 https://www.edsurge.com/news/2016-04-18-gradescope-raises-2-6m-to-apply-artificial-intelligence-to-grading-exams

301 http://www.wsj.com/articles/if-your-teacher-sounds-like-a-robot-you-might-be-on-to-something-1462546621

302 http://www.ravn.co.uk/

303 http://www.legalweek.com/legal-week/sponsored/2434504/is-artificial-intelligence-the-key-to-unlocking-innovation-in-your-law-firm

304 http://www.legalfutures.co.uk/latest-news/come-americans-legalzoom-gains-abs-licence

305 https://www.fairdocument.com/

306 http://msutoday.msu.edu/news/2014/using-data-to-predict-supreme-courts-decisions/

307 http://uk.businessinsider.com/ro-bots-may-make-legal-workers-obsolete-2015-8

308 http://www.telegraph.co.uk/news/2017/08/04/legal-robots-deployed-china-help-decide-thou-sands-cases/

309 https://www.bloomberg.com/news/arti-cles/2019-06-26/eqt-s-motherbrain-ready-to-take-on-46-billion-in-buyout-funds

310 https://www.sigfig.com/site/#/home

311 http://www.nytimes.com/2016/01/23/your-money/robo-advisers-for-investors-are-not-one-size-fits-all.html?_r=0

312 https://www.theguardian.com/business/2017/sep/06/deutsche-bank-boss-says-big-number-of-staff-will-lose-jobs-to-automation?CMP=share_btn_tw

313 https://www.thenational.ae/business/mashreq-to-shed-10-per-cent-of-headcount-in-next-12-months-as-artificial-intelligence-spending-pays-off-1.628437

314 http://www.businessinsider.com/r-technology-could-help-ubs-cut-workforce-by-30-percent-ceo-in-magazine-2017-10?IR=T

315 https://www.forbes.com/sites/bernard-marr/2019/02/15/the-revolutionary-way-of-using-artificial-intelligence-in-hedge-funds-the-case-of-aidyia/#392fcdc957ca

316 http://www.bloomberg.com/news/articles/2015-02-27/bridgewater-is-said-to-start-artificial-intelligence-team

317 http://www.wired.com/2016/01/the-rise-of-the-artificially-intelligent-hedge-fund/

318 https://next.ft.com/content/c31f8f44-033b-11e6-af1d-c47326021344 (Paywall)

319 https://www.bloomberg.com/news/features/2017-09-27/the-massive-hedge-fund-betting-on-ai

320 http://www.ft.com/cms/s/0/5eb91614-bee5-11e5-846f-79b0e3d20eaf.html#axzz3zEmSvuZs

321 https://www.bloomberg.com/news/features/2017-09-27/the-massive-hedge-fund-betting-on-ai

322 https://www.ft.com/content/16b8ffb6-7161-11e7-aca6-c6bd07df1a3c

323 http://uk.businessinsider.com/high-sala-

ry-jobs-will-be-automated-2016-3

324 http://www.fiercefinanceit.com/story/will-regu-latory-compliance-drive-artificial-intelligence-adop-tion/2016-01-05

325 http://www.liverpoolecho.co.uk/news/business/liverpool-fc-sponsor-standard-chartered-11104215

326 https://news.gallup.com/opinion/chair-man/212045/world-broken-workplace.aspx-?g_source=position1&g_medium=related&g_cam-paign=tiles

327 http://www.ft.com/cms/s/0/dfe218d6-9038-11e3-a776-00144feab7de.html#axzz3yUOe9Hkp

328 http://uk.businessinsider.com/googles-eric-schmidt-im-a-job-elimination-denier-on-the-risk-of-robots-stealing-jobs-2017-5

329 http://www.eastoftheweb.com/short-stories/UBooks/BoyCri.shtml

330 This useful phrase was originally coined by my friends at the Singularity Bros podcast, http://singu-laritybros.com/

331 http://blog.pmarca.com/2014/06/13/this-is-probably-a-good-time-to-say-that-i-dont-believe-ro-bots-will-eat-all-the-jobs/

332 http://uk.businessinsider.com/social-skills-becoming-more-important-as-robots-enter_workforce-2015-12

333 http://www.history.com/topics/inventions/automated-teller-machines

334 http://www.theatlantic.com/technology/archive/2015/03/a-brief-history-of-the-atm/388547/

335 http://www.wsj.com/articles/SB10001424052748704463504575301051844937276

336 http://kalw.org/post/robotic-seals-comfort-dementia-patients-raise-ethical-concerns#stream/0

337 http://www.scmp.com/week-asia/business/article/2104809/why-japan-will-profit-most-artificial-intelligence

338 http://viterbi.usc.edu/news/news/2013/a-virtual-therapist.htm

339 http://observer.com/2014/08/study-people-are-more-likely-to-open-up-to-a-talking-computer-than-a-human-therapist/

340 http://mindthehorizon.com/2015/09/21/avatar-virtual-reality-mental-health-tech/

341 http://www.handmadecake.co.uk/

342 http://www.bbc.co.uk/news/magazine-15551818

343 http://www.oxforddnb.com/view/article/19322

344 http://www.inc.com/john-brandon/22-inspiring-quotes-from-famous-entrepreneurs.html

345 https://www.edge.org/conversation/kevin_kelly-the-technium

346 https://quoteinvestigator.com/2011/11/05/computers-useless/#:~:text=The%20following%20mordant%20quotation%20has,can%20only%20give%20you%20answers.

347 http://www.ft.com/cms/s/2/c5cf07c4-bf8e-11e5-846f-79b0e3d20eaf.html#axzz3yLGlrr1J

348 http://www.bls.gov/cps/cpsaat11.htm

349 https://en.wikipedia.org/wiki/No_Man%27s_Sky

350 http://googleresearch.blogspot.co.uk/2015/06/inceptionism-going-deeper-into-neural.html

351 http://www.ft.com/cms/s/2/c5cf07c4-bf8e-11e5-846f-79b0e3d20eaf.html#axzz3yLGlrr1J

352 In case you only recently arrived on this planet, that was a reference to the sainted Douglas Adam's "Hitchhiker's Guide to the Galaxy" series. If you haven't read it, I recommend that you put this book down and read that one instead. I won't be offended. But please come back here afterwards.

353 http://philpapers.org/archive/DANHAT.pdf

354 The novel is sometimes said to have originated in the early 18th century, but in fact it is a much older art form. What happened then was that writers began publishing books which described life as they actually saw it. https://en.wikipedia.org/wiki/Novel#18th_century_novel

355 I am indebted to AGI researcher Randal Koene for this observation.

356 https://en.wikiquote.org/wiki/Bette_Davis

357 http://www.economist.com/node/17722567

358 http://quoteinvestigator.com/2011/11/16/robots-buy-cars/

359 http://thegreatdepressioncauses.com/unemployment/

360 http://www.statista.com/statistics/268830/unemployment-rate-in-eu-countries/

361 http://www.statista.com/statistics/266228/youth-unemployment-rate-in-eu-countries/

362 http://www.scottsantens.com/

363 http://www.economonitor.com/dolan-econ/2014/01/27/a-universal-basic-income-conservative-progressive-and-libertarian-perspectives-part-3-of-a-series/

364 https://www.reddit.com/r/BasicIncome/wiki/index#wiki_that.27s_all_very_well.2C_but_where.27s_the_evidence.3F

365 https://www.reddit.com/r/BasicIncome/wiki/studies

366 http://basicincome.org.uk/2013/08/health-forget-mincome-poverty/

367 http://fivethirtyeight.com/features/universal-basic-income/?utm_content=buffer71a7e&utm_medium=social&utm_source=plus.google.com&utm_campaign=buffer

368 http://www.fastcoexist.com/3052595/how-finlands-exciting-basic-income-experiment-will-work-and-what-we-can-learn-from-it

369 https://www.theguardian.com/society/2020/

may/07/finnish-basic-income-pilot-improved-well-being-study-finds-coronavirus

370 https://www.cnbc.com/2017/09/21/silicon-val-ley-giant-y-combinator-to-branch-out-basic-in-come-trial.html

371 https://www.wired.com/story/y-combinator-learns-basic-income-is-not-so-basic-after-all/

372 https://openresearchlab.org/basic-income/info/our-plan

373 https://en.wikipedia.org/wiki/Sodomy_laws_in_the_United_States#References

374 http://blogs.wsj.com/washwire/2015/03/09/sup-port-for-gay-marriage-hits-all-time-high-wsjnbc-news-poll/

375 http://www.huffingtonpost.com/2009/05/06/ma-jority-of-americans-wan_n_198196.html

376 https://www.businessinsider.com/legal-marijua-na-states-2018-1?r=US&IR=T

377 http://blogs.seattletimes.com/today/2013/08/washingtons-pot-law-wont-get-federal-challenge/

378 http://www.bbc.co.uk/news/magazine-35525566

379 https://medium.com/basic-income/wouldnt-un-conditional-basic-income-just-cause-massive-infla-tion-fe71d69f15e7#.3yezsngej

380 A splendidly cynical idea from my friend Matt Leach.

381 http://www.forbes.com/sites/greatspecula-tions/2012/12/05/how-i-know-higher-taxes-would-be-good-for-the-economy/#5b0c080b3ec1

382 http://taxfoundation.org/article/what-evidence-taxes-and-growth

383 https://en.wikipedia.org/wiki/Laffer_curve

384 http://www.bbc.co.uk/news/uk-poli-tics-26875420

385 A minor character in Shakespeare's Henry VI called Dick the Butcher has the memorable line, "First thing we do, let's kill all the lawyers." It seems Shakespeare was not fond of lawyers: http://www.spectacle.org/797/finkel.html

386 https://www.thersa.org/action-and-research/rsa-projects/public-services-and-communities-fold-er/basic-income/

387 http://www.icalculator.info/news/UK_average_earnings_2014.html

388 http://www.telegraph.co.uk/finance/economics/12037623/Paying-all-UK-citizens-155-a-week-may-be-an-idea-whose-time-has-come.html

389 https://medium.com/conversations-with-tyler/tyler-cowen-sam-altman-ai-tech-business-58f530417522

390 https://qz.com/911968/bill-gates-the-robot-that-takes-your-job-should-pay-taxes/

391 http://timharford.com/2016/05/could-an-income-for-all-provide-the-ultimate-safety-net/

392 http://archive.intereconomics.eu/year/2017/2/the-basics-of-basic-income/

393 http://fee.org/freeman/the-economic-fantasy-of-star-trek/

394 https://www.wired.co.uk/news/archive/2012-11/16/iain-m-banks-the-hydrogen-sonata-review

395 https://www.mckinsey.com/industries/retail/our-insights/automation-in-retail-an-executive-overview-for-getting-ready

396 https://www.forbes.com/sites/cognitiveworld/2019/12/04/review-of-more-from-less-by-an-

drew-mcafee/#1f5b537c130e

397 https://medium.com/@PeterDiaman-dis/a-bridge-to-abundance-6d83738d55dd

398 http://history.hanover.edu/courses/ex-cerpts/165acton.html

399 http://www.federalreserve.gov/econres-data/2014-economic-well-being-of-us-house-holds-in-2013-executive-summary.htm

400 http://www.theguardian.com/busi-ness/2016/jan/18/richest-62-billion-aires-wealthy-half-world-population-combined

401 http://www.bbc.co.uk/news/magazine-26613682

402 I'm indebted to Dr Justin Stewart, an investor, for prodding me to address the issue of assets more closely.

403 World Bank data: https://data.worldbank.org/indicator/NY.GDP.MKTP.CD?view=chart

404 https://en.wikipedia.org/wiki/Event_horizon

405 http://www.wired.com/2016/02/vr-moral-im-perative-or-opiate-of-masses/

406 http://motherboard.vice.com/read/sleep-tech-

will-widen-the-gap-between-the-rich-and-the-poor

407 https://en.wikipedia.org/wiki/Sex_and_drugs_and_rock_and_roll

408 https://www.cnbc.com/2017/08/09/initial-coin-offerings-surpass-early-stage-venture-capital-funding.html

409 http://www.telegraph.co.uk/technology/2017/08/24/bitcoin-price-stays-4000-will-continue-rise-will-bubble-burst/

410 http://www.dugcampbell.com/byzantine-generals-problem/

411 http://www.economistinsights.com/technology-innovation/analysis/money-no-middleman/tab/1

412 https://edge.org/conversation/john_markoff-the-next-wave

413 https://en.wikipedia.org/wiki/Milgram_experiment

414 http://www.prisonexp.org/

415 http://www.newyorker.com/magazine/2017/01/30/doomsday-prep-for-the-super-rich

416 http://fourhourworkweek.com/2014/08/29/kev-

in-kelly/

417 https://www.ted.com/speakers/hans_rosling

418 https://ourworldindata.org/

419 https://www.edge.org/conversation/kevin_kelly-the-technium

420 In the first and second editions of this book, these snapshots were of 2025, 2035, and 2045. An example of the first part of Amara's Law in action.

421 http://money.cnn.com/2015/06/23/investing/facebook-walmart-market-value/

422 http://lcfi.ac.uk/media/uploads/files/CFI_2017_programme.pdf

423 http://www.oxfordmartin.ox.ac.uk/research/programmes/tech-employment

424 The Machine Intelligence Research Institute (MIRI) in Northern California, The Future of Humanity Institute (FHI) and the Centre for the Study of Existential Risk (CSER) in England's Oxford and Cambridge respectively, and the Future of Life Institute (FLI) in Massachussetts.

425 There is a lot of useful insight into prediction markets here: https://bitedge.com/blog/prediction-

markets-are-about-to-be-a-big-deal/

426 Paul Boyer, 'Dr. Strangelove', a chapter in "Past Imperfect: History According to the Movies" edited by Mark C. Carnes

427 http://s05.static-shell.com/content/dam/shell/static/public/downloads/brochures/corporate-pkg/scenarios/explorers-guide.pdf

428 The Programme was established in January 2015 with funding from Citibank, one of the largest financial institutions in the world. The Oxford Martin school was set up as part of Oxford University in 2005, as an institution dedicated to understanding the threats and opportunities facing humanity in the 21st century. It is named after James Martin, a writer, consultant and entrepreneur, who founded the school with the largest donation ever made to the university – which was no mean feat given that Oxford was founded 1,000 years ago, and is the oldest university in the world (after Bologna in Italy).

429 https://www.youtube.com/watch?v=ou2Wzb-vgOJ8&feature=youtu.be&t=26472

430 This depends on the two planets being pretty much as close as they ever get.

431 http://fortune.com/2015/11/10/us-unemployment-rate-economy/

432 This and the other quotes in this paragraph and the next one are from Chapter 10: Toward a New Economic Paradigm.

433 Brynjolfsson is the director of the MIT Center for Digital Business and McAfee is a principal research scientist there.

434 The word "inequality" crops up 42 times in the book, including in the titles of sources, but the authors never explicitly connect it with "spread".

435 The loosely-organised protest organisation that sprang up after the 2008 credit crunch to campaign against inequality.

436 Chapter 12: Learning to Race with the Machines: Recommendations for Individuals.

437 Chapter 13: Policy Recommendations.

438 Chapter 14: Long-Term Recommendations.

439 https://www.wired.com/2017/08/robots-will-not-take-your-job

440 https://www.forbes.com/sites/cognitive-world/2019/12/04/review-of-more-from-less-by-andrew-mcafee/#3c69db69130e

441 http://www.susskind.com/

442 https://www.forbes.com/sites/cognitive-world/2020/01/30/a-world-without-work-by-daniel-susskind-a-book-review/#17db4a396dd7

443 http://www.scottsantens.com/

444 https://www.reddit.com/r/BasicIncome/ and https://www.reddit.com/r/basicincome/wiki/index

445 https://www.youtube.com/watch?v=7Pq-S557X-QU.

446 https://www.youtube.com/watch?v=C5MVX-dg6nho.

447 http://www.bankofengland.co.uk/publications/Pages/speeches/2015/864.aspx

448 https://newrepublic.com/article/69326/call-the-wolf

449 http://www.ft.com/cms/s/0/dfe218d6-9038-11e3-a776-00144feab7de.html#axzz3stkJb1V2

450 http://www.socialeurope.eu/2015/10/the-limits-of-the-digital-revolution-why-our-washing-machines-wont-go-to-the-moon/

451 https://www.aeaweb.org/articles.php?-

doi=10.1257/jep.29.3.3

452 https://reason.com/archives/2015/03/03/how-to-survive-a-robot-uprisin

453 http://www.politico.com/magazine/story/2013/11/the-robots-are-here-098995

454 http://www.forbes.com/sites/danschawbel/2015/08/04/geoff-colvin-why-humans-will-triumph-over-machines/2/

455 https://www.forbes.com/sites/cognitive-world/2019/10/22/a-brexiteer-among-the-robotsa-review-of-the-ai-economy-by-roger-bootle/#-66951c725ea3

Made in the USA
Columbia, SC
27 February 2021

33660117R00265